T0201283

JET SINGLE-TIME
LAGRANGE GEOMETRY
AND ITS APPLICATIONS

JET SINGLE-TIME LAGRANGE GEOMETRY AND ITS APPLICATIONS

Vladimir Balan
University Politehnica of Bucharest

Mircea Neagu
University Transilvania of Braşov

A JOHN WILEY & SONS, INC., PUBLICATION

Published by John Wiley & Sons, Inc., Hoboken, New Jersey
Published simultaneously in Canada

For general information on our other products and services or for technical support, please contact our Customer Care Department within the United States at (800) 762-2974, outside the United States at (317) 572-3993 or fax (317) 572-4002.

Wiley also publishes its books in a variety of electronic formats. Some content that appears in print may not be available in electronic formats. For more information about Wiley products, visit our web site at www.wiley.com.

Library of Congress Cataloging-in-Publication Data is available.

ISBN 978-1-118-12755-1

Printed in the United States of America.

10 9 8 7 6 5 4 3 2 1

CONTENTS

PREFACE

The 1-jet fiber bundle is a basic object in the study of classical and quantum field theories ([77], [8], etc.). For this a reason, a lot of authors (Asanov [8], Martínez [51], Saunders [89], Vondra [98], [99], Văcaru [96], [97] and many others) studied the differential geometry of the 1-jet spaces. Considering the geometrical studies of Asanov [8] and using as a pattern the Lagrangian geometrical ideas developed by Miron, Anastasiei, or Bucătaru in the monographs [55] and [24], the second author of this book has recently developed the *Riemann-Lagrange geometry of 1-jet spaces* [65], which is very suitable for the geometrical study of the *relativistic time-dependent* (*rheonomic* or *non-autonomous*) *Lagrangians*. In this framework, we refer to the geometrization of Lagrangians depending on the usual *relativistic time* [67] or of Lagrangians depending on *relativistic multi-time* [65], [70].

It is important to note that a *classical time-dependent (rheonomic) Lagrangian geometry* (i.e. a geometrization of the Lagrangians depending on the usual *absolute time*) was sketched by Miron and Anastasiei at the end of the book [55] and accordingly developed by Anastasiei and Kawaguchi [1] or Frigioiu [34]. For such a reason, we shall further describe the main geometrical and physical aspects which differentiate the two geometrical theories: the *jet relativistic time-dependent Lagrangian geometry* [67] and the *classical time-dependent Lagrangian geometry* [55].

In this direction, we point out that the *relativistic time-dependent Lagrangian geometry* [67] naturally relies on the 1-jet space $J^1(\mathbb{R}, M)$, where $\mathbb{R}$ is the manifold of real numbers having the *temporal* coordinate t. This represents the usual *relativistic time*. We recall that the 1-jet space $J^1(\mathbb{R}, M)$ is regarded as a vector bundle over the product manifold $\mathbb{R} \times M$, having the fiber type $\mathbb{R}^n$, where n is the dimension of the *spatial* manifold M. In mechanical terms, if the manifold M has the *spatial* local coordinates $(x^i)_{i=\overline{1,n}}$, then the 1-jet vector bundle

$$J^1(\mathbb{R}, M) \to \mathbb{R} \times M \tag{0.1}$$

can be regarded as a *bundle of configurations* whose total space has the local coordinates (t, x^i, y^i_1); these transform by the rules [67]

$$\begin{cases} \widetilde{t} = \widetilde{t}(t), \\ \widetilde{x}^i = \widetilde{x}^i(x^j), \\ \widetilde{y}^i_1 = \dfrac{\partial \widetilde{x}^i}{\partial x^j} \dfrac{dt}{d\widetilde{t}} \cdot y^j_1. \end{cases} \tag{0.2}$$

We remark that the form of the jet transformation group defined by the rules (0.2) stands out for the *relativistic character* of the *time* t.

Comparatively, denoting by TM the tangent bundle of the *spatial* manifold M, we note that in the *classical time-dependent Lagrangian geometry* the *bundle of configurations* is the vector bundle [55]

$$\mathbb{R} \times TM \to M, \tag{0.3}$$

whose local coordinates (t, x^i, y^i) transform by the rules

$$\begin{cases} \widetilde{t} = t, \\ \widetilde{x}^i = \widetilde{x}^i(x^j), \\ \widetilde{y}^i = \dfrac{\partial \widetilde{x}^i}{\partial x^j} \cdot y^j. \end{cases} \tag{0.4}$$

Remark 1. The form of the transformation group (0.4) stands out for the *absolute character* of the *time* t.

We emphasize that the jet transformation group (0.2) from the *relativistic time-dependent Lagrangian geometry* is more general and more natural than the transformation group (0.4) used in the *classical time-dependent Lagrangian geometry*. This is because the last one ignores the temporal reparametrizations, emphasizing in this way the absolute character of the usual time coordinate t. Or, physically speaking, the relativity of time is an well-known fact.

From a geometrical point of view, we point out that the entire *classical time-dependent (rheonomic or non-autonomous) Lagrangian geometry* of Miron and Anastasiei [55] relies on the study of the *energy action functional*

$$\mathbb{E}_1(c) = \int_a^b L(t, x^i, y^i) dt,$$

where $L : \mathbb{R} \times TM \rightarrow \mathbb{R}$ is a Lagrangian function and $y^i = dx^i/dt$, whose Euler-Lagrange equations

$$\ddot{x}^i + 2G^i(t, x^i, y^i) = 0$$

produce the semispray G^i and the nonlinear connection $N^i_j = \partial G^i/\partial y^j$. In the sequel, the authors construct the adapted bases of vector and covector fields, together with the adapted components of the N-linear connections and their corresponding torsions and curvatures. But, because $L(t, x^i, y^i)$ is a real function, we deduce that the previous geometrical theory has the impediment that *the energy action functional depends on the reparametrizations $t \longleftrightarrow \tilde{t}$ of the same curve c.*

For example, in order to avoid this inconvenience, the Finsler case imposes the 1-positive homogeneity condition

$$L(t, x^i, \lambda y^i) = \lambda L(t, x^i, y^i), \ \forall \, \lambda > 0.$$

Alternatively, the *relativistic time-dependent Lagrangian geometry* from this book (see also [67]) uses the *relativistic energy action functional*

$$\mathbb{E}_2(c) = \int_a^b L(t, x^i, y_1^i) \sqrt{h_{11}(t)} dt,$$

where $L : J^1(\mathbb{R}, M) \rightarrow \mathbb{R}$ is a jet Lagrangian function and $h_{11}(t)$ is a Riemannian metric on the time manifold $\mathbb{R}$. This functional is now independent by the reparametrizations $t \longleftrightarrow \tilde{t}$ of the same curve c and the corresponding Euler-Lagrange equations take the form

$$\ddot{x}^i + 2H^{(i)}_{(1)1}\left(t, x^k, y_1^k\right) + 2G^{(i)}_{(1)1}\left(t, x^k, y_1^k\right) = 0,$$

where the coefficients $H^{(i)}_{(1)1}$, respectively $G^{(i)}_{(1)1}$, represent a temporal, respectively spatial, semispray. These semisprays produce a jet canonical nonlinear connection

$$\Gamma = \left(M^{(j)}_{(1)1} = 2H^{(j)}_{(1)1}, \quad N^{(j)}_{(1)k} = \frac{\partial G^{(j)}_{(1)1}}{\partial y_1^k} \right).$$

We further describe the local adapted components of connections, torsion, and curvature. We emphasize that the local adapted components of the jet geometrical objects involved in the present study obey the formalism used in the works [55], [24], [1], [34], [65], [67], [12].

In this respect, the authors of this book believe that the relativistic geometrical approach proposed in this monograph has more geometrical and physical meanings than the theory proposed by Miron and Anastasiei in [55].

As a final remark, we point out that for numerous mathematicians (such as Crampin [31], Krupková [45], [46], de León [48], Sarlet [88], Saunders [89], and others) the time-dependent Lagrangian geometry is constructed on the first jet bundle $J^1\pi$ of a fibered manifold $\pi : M^{n+1} \longrightarrow \mathbb{R}$. In their papers, if (t, x^i) are the local coordinates on the $n + 1$-dimensional manifold M such that t is a global coordinate for the fibers of the submersion π and x^i are transverse coordinates of the induced foliation, then a change of coordinates on M is given by

$$
\begin{cases}
\widetilde{t} = \widetilde{t}(t), & \dfrac{d\widetilde{t}}{dt} \neq 0, \\[2mm]
\widetilde{x}^i = \widetilde{x}^i(x^j, t), & \text{rank}\left(\dfrac{\partial \widetilde{x}^i}{\partial x^j}\right) = n.
\end{cases}
\tag{0.5}
$$

Although the 1-jet extension of the transformation rules (0.5) is more general than the transformation group (0.2), the authors of this book consider that the transformation group (0.2) is more appropriate for their final purpose: the development of the *relativistic time-dependent Lagrangian geometrical field theories*.

For example, in our monograph, starting with a non-degenerate Lagrangian function $L : J^1(\mathbb{R}, M) \to \mathbb{R}$ and an *a priori* given Riemannian metric $h_{11}(t)$ on the relativistic temporal manifold $\mathbb{R}$ (these geometrical objects produce together the Lagrangian $\mathcal{L} = L\sqrt{h_{11}(t)}$), one introduces the jet single-time gravitational potential

$$
\mathbb{G} = h_{11}dt \otimes dt + g_{ij}dx^i \otimes dx^j + h^{11}(t)g_{ij}(t, x, y)\delta y_1^i \otimes \delta y_1^j,
\tag{0.6}
$$

where

$$
g_{ij}(t, x, y) = \frac{h_{11}(t)}{2}\frac{\partial^2 L}{\partial y_1^i \partial y_1^j}, \qquad \delta y_1^i = dy_1^i + M_{(1)1}^{(i)}dt + N_{(1)j}^{(i)}dx^j.
$$

Note that the above jet single-time gravitational potential $\mathbb{G}$ is a global geometrical object on $J^1(\mathbb{R}, M)$, with respect to the group of transformations (0.2). Moreover, it is characterized (as in the Miron and Anastasiei case [55]) by some natural *geometrical Einstein equations* [67]. These geometrical Einstein equations will be described in the next chapters of this book.

At the same time, the transformation group (0.2) is more appropriate for the development of a *relativistic time-dependent Lagrangian electromagnetic theory*, whose *jet single-time electromagnetic field* is defined by

$$
\mathbb{F} = F_{(i)j}^{(1)}\delta y_1^i \wedge dx^j,
$$

where

$$
F_{(i)j}^{(1)} = \frac{1}{2}\left[D_{(i)j}^{(1)} - D_{(j)i}^{(1)}\right],
$$

the *metrical deflection d-tensors* $D_{(i)j}^{(1)} = h^{11}g_{im}y_{1|j}^m$ being produced only by the jet Lagrangian $\mathcal{L} = L\sqrt{h_{11}(t)}$, via its Cartan canonical Γ-linear connection. From such a perspective, the electromagnetic components $F_{(i)j}^{(1)}$ are governed by some natural *geometrical Maxwell equations*. These geometrical Maxwell equations will be also presented in this book and they naturally generalize the already classical Maxwell equations from Miron and Anastasiei's theory [55], which has many applications in Theoretical Physics, such as *Electrodynamics*, *Relativistic Optics*, or *Relativity* and *Electromagnetism*.

In our book, we will show that our jet single-time Lagrangian geometry also gives a lot of applications to various domains of sciences: *Theoretical Physics* (the gravitational theory produced by the Berwald-Moór metric or by the Chernov metric), *Atmospheric Physics*, *Economy*, or *Theoretical Biology*. As well, at the end of the book, there are presented the basic elements of the Kosambi-Cartan-Chern theory on the 1-jet space $J^1(\mathbb{R}, M)$, which extend the approach developed in [44], [26], [27], [24], [4].

Finally, the authors of this book express their gratitude to Professors R. Miron, M. Anastasiei, Gh. Atanasiu, C. Udrişte, D. Opriş, D. G. Pavlov, M. Rahula, P.C. Stavrinos, K. Teleman, K. Trenčevski, V. Obădeanu, Gh. Pitiş, Gh. Munteanu, I. Mihai, V. Prepeliţă, G. Pripoae, M. Crâşmăreanu, E. Păltânea, I. R. Nicola, M. Postolache, M. Lupu, E. Stoica, M. Păun and C. Radu for stimulative discussions on the geometrical methods used in the applicative research from this book. Special thanks are addressed to the late Professor R. G. Beil who kindly provided us G. S. Asanov's essential paper [8], which settles on 1-jet spaces the geometrical foundations for a similar related framework.

Vladimir Balan
Mircea Neagu
April 21, 2011

PART I

THE JET SINGLE-TIME LAGRANGE GEOMETRY

CHAPTER 1

JET GEOMETRICAL OBJECTS DEPENDING ON A RELATIVISTIC TIME

The differential geometry of the 1-jet space $J^1(\mathbb{R}, M)$ was intensively studied by numerous authors: Crampin [31], Krupková [46], de León [48], Martínez [51], Sarlet [88], Saunders [89], Vondra [98], [99], etc. Compared to their approaches, our framework focuses on the local decomposition of the involved geometrical objects relative to adapted bases associated to a given nonlinear connection. In the present Chapter, developing further the geometrical studies initiated by Asanov [8] and using as a pattern the geometrical ideas developed by Miron and Anastasiei in [55], we study a collection of jet geometrical objects (*d-tensors, relativistic time-dependent semis-prays, harmonic curves,* and *nonlinear connections*), together with the underlying fundamental geometrical relations which relate them. These geometrical concepts are essential for the subsequent construction of the geometrization (in the sense of R. Miron) of the 1-jet space $J^1(\mathbb{R}, M)$. This geometrization on the 1-jet spaces will therefore assume the decomposition of the geometrical objects involved into their local adapted components.

Jet Single-Time Lagrange Geometry and Its Applications   **3**
1-st Edition. By Vladimir Balan and Mircea Neagu.
© 2011 John Wiley & Sons, Inc. Published 2011 John Wiley & Sons, Inc.

1.1   d-TENSORS ON THE 1-JET SPACE $J^1(\mathbb{R}, M)$

Let us consider the 1-jet fiber bundle

$$J^1(\mathbb{R}, M) \to \mathbb{R} \times M, \tag{1.1}$$

whose local coordinates (t, x^i, y_1^i) transform by the rules

$$\begin{cases} \widetilde{t} = \widetilde{t}(t), \\[2mm] \widetilde{x}^i = \widetilde{x}^i(x^j), \\[2mm] \widetilde{y}_1^i = \dfrac{\partial \widetilde{x}^i}{\partial x^j} \dfrac{dt}{d\widetilde{t}} \cdot y_1^j. \end{cases} \tag{1.2}$$

It is well known that in the study of the geometry of a fiber bundle an important role is played by tensors. For such a reason, let us consider

$$\left\{ \frac{\partial}{\partial t}, \frac{\partial}{\partial x^i}, \frac{\partial}{\partial y_1^i} \right\} \subset \mathcal{X}(J^1(\mathbb{R}, M)),$$

the canonical basis of vector fields on the 1-jet space $J^1(\mathbb{R}, M)$, together with its dual basis of 1-forms

$$\left\{ dt, dx^i, dy_1^i \right\} \subset \mathcal{X}^*(J^1(\mathbb{R}, M)).$$

In this context, let us remark that, doing a transformation of jet local coordinates (1.2), the following transformation rules hold true:

$$\begin{cases} \dfrac{\partial}{\partial t} = \dfrac{d\widetilde{t}}{dt} \dfrac{\partial}{\partial \widetilde{t}} + \dfrac{\partial \widetilde{y}_1^j}{\partial t} \dfrac{\partial}{\partial \widetilde{y}_1^j}, \\[4mm] \dfrac{\partial}{\partial x^i} = \dfrac{\partial \widetilde{x}^j}{\partial x^i} \dfrac{\partial}{\partial \widetilde{x}^j} + \dfrac{\partial \widetilde{y}_1^j}{\partial x^i} \dfrac{\partial}{\partial \widetilde{y}_1^j}, \\[4mm] \dfrac{\partial}{\partial y_1^i} = \dfrac{\partial \widetilde{x}^j}{\partial x^i} \dfrac{dt}{d\widetilde{t}} \dfrac{\partial}{\partial \widetilde{y}_1^j}, \end{cases} \tag{1.3}$$

and

$$\begin{cases} dt = \dfrac{dt}{d\widetilde{t}} d\widetilde{t}, \\[4mm] dx^i = \dfrac{\partial x^i}{\partial \widetilde{x}^j} d\widetilde{x}^j, \\[4mm] dy_1^i = \dfrac{\partial y_1^i}{\partial \widetilde{t}} d\widetilde{t} + \dfrac{\partial y_1^i}{\partial \widetilde{x}^j} d\widetilde{x}^j + \dfrac{\partial x^i}{\partial \widetilde{x}^j} \dfrac{d\widetilde{t}}{dt} d\widetilde{y}_1^j. \end{cases} \tag{1.4}$$

Taking into account that the transformation rules (1.3) and (1.4) lead to complicated transformation rules for the components of classical tensors on the 1-jet space $J^1(\mathbb{R}, M)$, we consider that in the geometrical study of the 1-jet fiber bundle $J^1(\mathbb{R}, M)$ a central role is played by the *distinguished tensors* (*d-tensors*).

Definition 2. A geometrical object $D = \left(D_{1k(1)(l)\ldots}^{1i(j)(1)\ldots} \right)$ on the 1-jet vector bundle $J^1(\mathbb{R}, M)$, whose local components transform by the rules

$$D_{1k(1)(l)\ldots}^{1i(j)(1)\ldots} = \widetilde{D}_{1r(1)(s)\ldots}^{1p(m)(1)\ldots} \frac{dt}{d\widetilde{t}} \frac{\partial x^i}{\partial \widetilde{x}^p} \left(\frac{\partial x^j}{\partial \widetilde{x}^m} \frac{d\widetilde{t}}{dt} \right) \frac{d\widetilde{t}}{dt} \frac{\partial \widetilde{x}^r}{\partial x^k} \left(\frac{\partial \widetilde{x}^s}{\partial x^l} \frac{dt}{d\widetilde{t}} \right) \ldots, \qquad (1.5)$$

is called a *d-tensor field*.

Remark 3. The utilization of parentheses for certain indices of the local components $D_{1k(1)(l)\ldots}^{1i(j)(1)\ldots}$ of the distinguished tensor D on $J^1(\mathbb{R}, M)$ will be rigorously motivated after the introduction of the geometrical concept of *nonlinear connection* on the 1-jet space $J^1(\mathbb{R}, M)$. For the moment, we point out that the pair of indices " $\genfrac{}{}{0pt}{}{(j)}{(1)}$ " or " $\genfrac{}{}{0pt}{}{(1)}{(l)}$ " behaves like a single index.

Remark 4. From a physical point of view, a d-tensor field D on the 1-jet vector bundle $J^1(\mathbb{R}, M) \to \mathbb{R} \times M$ can be regarded as a physical object defined on the *space of events* $\mathbb{R} \times M$, which is dependent by the *direction* or the *relativistic velocity* $y = (y_1^i)$. Such a perspective is intimately connected with the physical concept of *anisotropy*.

■ **EXAMPLE 1.1**

Let $L : J^1(\mathbb{R}, M) \to \mathbb{R}$ be a *relativistic time-dependent Lagrangian function*, where

$$J^1(\mathbb{R}, M) \ni (t, x^i, y_1^i) \to L(t, x^i, y_1^i) \in \mathbb{R}.$$

Then, the geometrical object $\mathbf{G} = \left(G_{(i)(j)}^{(1)(1)} \right)$, where

$$G_{(i)(j)}^{(1)(1)} = \frac{1}{2} \frac{\partial^2 L}{\partial y_1^i \partial y_1^j},$$

is a d-tensor field on $J^1(\mathbb{R}, M)$, which is called the *fundamental metrical d-tensor* produced by the jet Lagrangian function L. Note that the d-tensor field

$$G_{(i)(j)}^{(1)(1)}(t, x^i, y_1^i)$$

is a natural generalization for the metrical d-tensor field $g_{ij}(t, x^i, y^i)$ of a classical time-dependent (rheonomic) Lagrange space [55]

$$RL^n = (M, L(t, x^i, y^i)),$$

where $L : \mathbb{R} \times TM \to \mathbb{R}$ is an absolute time-dependent Lagrangian function.

■ **EXAMPLE 1.2**

The geometrical object $\mathbf{C} = \left(\mathbf{C}^{(i)}_{(1)} \right)$, where $\mathbf{C}^{(i)}_{(1)} = y^i_1$, represents a d-tensor field on the 1-jet space $J^1(\mathbb{R}, M)$. This is called the *canonical Liouville d-tensor field* of the 1-jet vector bundle $J^1(\mathbb{R}, M)$. Remark that the d-tensor field $\mathbf{C}$ naturally generalizes the classical Liouville vector field [55]

$$\mathbb{C} = y^i \frac{\partial}{\partial y^i},$$

used in the Lagrangian geometry of the tangent bundle TM.

■ **EXAMPLE 1.3**

Let $h = (h_{11}(t))$ be a Riemannian metric on the relativistic time axis $\mathbb{R}$ and let us consider the geometrical object $\mathbf{J}_h = \left(J^{(i)}_{(1)1j} \right)$, where

$$J^{(i)}_{(1)1j} = h_{11} \delta^i_j,$$

δ^i_j being the Kronecker symbol. Then, the geometrical object $\mathbf{J}_h$ is a d-tensor field on $J^1(\mathbb{R}, M)$, which is called the *h-normalization d-tensor field* of the 1-jet space $J^1(\mathbb{R}, M)$. We underline that our h-normalization d-tensor field $\mathbf{J}_h$ of the 1-jet space $J^1(\mathbb{R}, M)$ naturally generalizes the *tangent structure* [55]

$$\mathbb{J} = \delta^i_j \frac{\partial}{\partial y^i} \otimes dx^j = \frac{\partial}{\partial y^i} \otimes dx^i,$$

constructed in the Lagrangian geometry of the tangent bundle TM.

■ **EXAMPLE 1.4**

Using preceding notations, we consider the set of local functions $\mathbf{L}_h = \left(L^{(i)}_{(1)11} \right)$, where

$$L^{(i)}_{(1)11} = h_{11} y^i_1.$$

The geometrical object $\mathbf{L}_h$ is a d-tensor field on $J^1(\mathbb{R}, M)$, which is called the *h-canonical Liouville d-tensor field* of the 1-jet space $J^1(\mathbb{R}, M)$.

1.2 RELATIVISTIC TIME-DEPENDENT SEMISPRAYS. HARMONIC CURVES

We point out that the notions of *d-tensor* and *classical tensor* on the 1-jet space $J^1(\mathbb{R}, M)$ are distinct ones. However, we will show after the introduction of the geometrical concept of *nonlinear connection*, that any d-tensor is a classical tensor on $J^1(\mathbb{R}, M)$. Conversely, this statement is not true. For instance, we construct in the sequel two classical global tensors which are not d-tensors on $J^1(\mathbb{R}, M)$. We talk about two geometrical notions: the *temporal semispray* and the *spatial semispray* on $J^1(\mathbb{R}, M)$. These concepts allow us afterward to introduce the geometrical concept of *relativistic time-dependent semispray* on the 1-jet space $J^1(\mathbb{R}, M)$.

Definition 5. A global tensor H on the 1-jet space $J^1(\mathbb{R}, M)$, which is locally expressed by

$$H = dt \otimes \frac{\partial}{\partial t} - 2H^{(j)}_{(1)1} dt \otimes \frac{\partial}{\partial y^j_1}, \tag{1.6}$$

is called a *temporal semispray* on $J^1(\mathbb{R}, M)$.

Taking into account that the temporal semispray H is a global classical tensor on $J^1(\mathbb{R}, M)$, by direct local computations, we find the following:

Proposition 6. (i) *The local components* $H^{(j)}_{(1)1}$ *of the temporal semispray H transform by the rules*

$$2\widetilde{H}^{(k)}_{(1)1} = 2H^{(j)}_{(1)1} \left(\frac{dt}{d\widetilde{t}}\right)^2 \frac{\partial \widetilde{x}^k}{\partial x^j} - \frac{dt}{d\widetilde{t}} \frac{\partial \widetilde{y}^k_1}{\partial t}. \tag{1.7}$$

(ii) *Conversely, to give a temporal semispray on $J^1(\mathbb{R}, M)$ is equivalent to giving a set of local functions* $H = \left(H^{(j)}_{(1)1}\right)$ *which transform by the rules (1.7).*

■ **EXAMPLE 1.5**

Let us consider that $h = (h_{11}(t))$ is a Riemannian metric on the temporal manifold $\mathbb{R}$ and let

$$\kappa^1_{11} = \frac{h^{11}}{2} \frac{dh_{11}}{dt},$$

where $h^{11} = 1/h_{11} > 0$, be its Christoffel symbol. Taking into account that we have the transformation rule

$$\widetilde{\kappa}^1_{11} = \kappa^1_{11} \frac{dt}{d\widetilde{t}} + \frac{d\widetilde{t}}{dt} \frac{d^2t}{d\widetilde{t}^2}, \tag{1.8}$$

we deduce that the local components

$$\overset{\circ}{H}^{(j)}_{(1)1} = -\frac{1}{2} \kappa^1_{11} y^j_1$$

define a temporal semispray $\overset{\circ}{H} = \left(\overset{\circ}{H}^{(j)}_{(1)1}\right)$ on $J^1(\mathbb{R}, M)$. This is called the *canonical temporal semispray associated to the temporal metric $h_{11}(t)$.*

Definition 7. A global tensor G on the 1-jet space $J^1(\mathbb{R}, M)$, which is locally expressed by

$$G = y_1^i dt \otimes \frac{\partial}{\partial x^i} - 2G_{(1)1}^{(j)} dt \otimes \frac{\partial}{\partial y_1^j}, \tag{1.9}$$

is called a *spatial semispray* on $J^1(\mathbb{R}, M)$.

As in the case of the temporal semispray, by direct local computations, we can prove without difficulties the following statements:

Proposition 8. (i) *The local components* $G_{(1)1}^{(j)}$ *of the spatial semispray G transform by the rules*

$$2\widetilde{G}_{(1)1}^{(k)} = 2G_{(1)1}^{(j)} \left(\frac{dt}{d\widetilde{t}}\right)^2 \frac{\partial \widetilde{x}^k}{\partial x^j} - \frac{\partial x^i}{\partial \widetilde{x}^j} \frac{\partial \widetilde{y}_1^k}{\partial x^i} \widetilde{y}_1^j. \tag{1.10}$$

(ii) *Conversely, to give a spatial semispray on $J^1(\mathbb{R}, M)$ is equivalent to giving a set of local functions* $G = \left(G_{(1)1}^{(j)}\right)$ *which transform by the rules (1.10).*

■ **EXAMPLE 1.6**

Let us consider that $\varphi = (\varphi_{ij}(x))$ is a semi-Riemannian metric on the spatial manifold M and let

$$\gamma_{jk}^i = \frac{\varphi^{im}}{2} \left(\frac{\partial \varphi_{jm}}{\partial x^k} + \frac{\partial \varphi_{km}}{\partial x^j} - \frac{\partial \varphi_{jk}}{\partial x^m}\right)$$

be its Christoffel symbols. Taking into account that we have the transformation rules

$$\widetilde{\gamma}_{qr}^p = \gamma_{jk}^i \frac{\partial \widetilde{x}^p}{\partial x^i} \frac{\partial x^j}{\partial \widetilde{x}^q} \frac{\partial x^k}{\partial \widetilde{x}^r} + \frac{\partial \widetilde{x}^p}{\partial x^l} \frac{\partial^2 x^l}{\partial \widetilde{x}^q \partial \widetilde{x}^r}, \tag{1.11}$$

we deduce that the local components

$$\mathring{G}_{(1)1}^{(j)} = \frac{1}{2}\gamma_{kl}^j y_1^k y_1^l$$

define a spatial semispray $\mathring{G} = \left(\mathring{G}_{(1)1}^{(j)}\right)$ on $J^1(\mathbb{R}, M)$. This is called the *canonical spatial semispray associated to the spatial metric* $\varphi_{ij}(x)$.

Remark 9. It is important to note that our notions of temporal and spatial semispray naturally generalize the classical notion of semispray (or semigerbe in the French terminology) which was defined since 1960's (Dazord, Klein, Foulon, de León, Miron, and Anastasiei, etc.) as a global vector field. Comparatively, we point out that our temporal or spatial semisprays can be regarded in the form

$$H = dt \otimes H_1, \qquad G = dt \otimes G_1.$$

Obviously, the geometrical objects (similarly with the classical concepts of semisprays or semigerbes)

$$H_1 = \frac{\partial}{\partial t} - 2H_{(1)1}^{(j)} \frac{\partial}{\partial y_1^j}$$

and

$$G_1 = y_1^i \frac{\partial}{\partial x^i} - 2G_{(1)1}^{(j)} \frac{\partial}{\partial y_1^j}$$

cannot be regarded as global vector fields on $J^1(\mathbb{R}, M)$ because they behave as the components of some d-covector fields on the 1-jet space $J^1(\mathbb{R}, M)$. In other words, taking into account the transformation rules (1.2), the geometrical objects H_1 and G_1 transform by the laws

$$\widetilde{H}_1 = \frac{dt}{d\widetilde{t}} H_1, \qquad \widetilde{G}_1 = \frac{dt}{d\widetilde{t}} G_1.$$

In conclusion, if we work only with particular transformations (1.2) in which the time t is absolute one (i.e., $\widetilde{t} = t$), then the geometrical objects H_1 and G_1 become global vector fields and, consequently, we recover the classical definition of a semispray or a semigerbe.

Definition 10. A pair $\mathcal{S} = (H, G)$, which consists of a temporal semispray H and a spatial semispray G, is called a *relativistic time-dependent semispray* on the 1-jet space $J^1(\mathbb{R}, M)$.

Remark 11. The geometrical concept of *relativistic time-dependent semispray* on the 1-jet space $J^1(\mathbb{R}, M)$ naturally generalizes the already classical notion of *time-dependent semispray* on $\mathbb{R} \times TM$, used by Miron and Anastasiei in [55].

■ **EXAMPLE 1.7**

The pair $\overset{\circ}{\mathcal{S}} = (\overset{\circ}{H}, \overset{\circ}{G})$, where $\overset{\circ}{H}$ [respectively $\overset{\circ}{G}$] is the canonical temporal (respectively spatial) semispray associated to the temporal (respectively spatial) metric $h_{11}(t)$ [respectively $\varphi_{ij}(x)$], is a relativistic time-dependent semispray on the 1-jet space $J^1(\mathbb{R}, M)$. This is called the *canonical relativistic time-dependent semispray associated to the pair of metrics* $(h_{11}(t), \varphi_{ij}(x))$.

In order to underline the importance of the canonical relativistic time-dependent semispray $\overset{\circ}{\mathcal{S}}$ associated to the pair of metrics $(h_{11}(t), \varphi_{ij}(x))$, we give the following geometrical result which characterizes the relativistic time-dependent semisprays on 1-jet spaces:

Proposition 12. *Let $(\mathbb{R}, h_{11}(t))$ be a Riemannian manifold and let $(M, \varphi_{ij}(x))$ be a semi-Riemannian manifold. Let $\mathcal{S} = (H, G)$ be an arbitrary relativistic time-dependent semispray on the 1-jet space $J^1(\mathbb{R}, M)$. Then, there exists on $J^1(\mathbb{R}, M)$ a unique pair of d-tensors*

$$\mathcal{T} = \left(T_{(1)1}^{(i)}, S_{(1)1}^{(i)} \right),$$

such that

$$S = \overset{\circ}{S} - T,$$

where $\overset{\circ}{S} = (\overset{\circ}{H}, \overset{\circ}{G})$ is the canonical relativistic time-dependent semispray associated to the pair of metrics $(h_{11}(t), \varphi_{ij}(x))$.

Proof: Taking into account that the difference between two temporal (respectively spatial) semisprays is a d-tensor [see the relations (1.7) and (1.10)], we find the required result.

Now, let us fix on the 1-jet space $J^1(\mathbb{R}, M)$ an arbitrary relativistic time-dependent semispray

$$S = (H, G) = \left(H^{(i)}_{(1)1}(t, x^k, y_1^k), G^{(i)}_{(1)1}(t, x^k, y_1^k) \right).$$

Definition 13. A smooth curve $c : t \in I \subset \mathbb{R} \to c(t) = (x^i(t)) \in M$, which verifies the second-order differential equations (SODEs)

$$\frac{d^2 x^i}{dt^2} + 2H^{(i)}_{(1)1}\left(t, x^k(t), \frac{dx^k}{dt} \right) + 2G^{(i)}_{(1)1}\left(t, x^k(t), \frac{dx^k}{dt} \right) = 0, \tag{1.12}$$

where i runs from 1 to n, is called a harmonic curve of the relativistic time-dependent semispray $S = (H, G)$.

Remark 14. The SODEs (1.12) are invariant under a transformation of coordinates given by (1.2). It follows that the form of equations (1.12), which give the harmonic curves of a relativistic time-dependent semispray $S = (H, G)$, have a global character on the 1-jet space $J^1(\mathbb{R}, M)$.

Remark 15. The equations of the *harmonic curves* (1.12) naturally generalize the equations of the *paths* of a time-dependent semispray from *classical rheonomic (non-autonomous) Lagrangian geometry* [55].

■ **EXAMPLE 1.8**

The equations of the harmonic curves of the canonical relativistic time-dependent semispray $\overset{\circ}{S} = (\overset{\circ}{H}, \overset{\circ}{G})$ associated to the pair of metrics $(h_{11}(t) , \varphi_{ij}(x))$ are

$$\frac{d^2 x^i}{dt^2} - \kappa_{11}^1(t)\frac{dx^i}{dt} + \gamma_{jk}^i(x)\frac{dx^j}{dt}\frac{dx^k}{dt} = 0. \tag{1.13}$$

These are the equations of the *affine maps* between the Riemannian manifold $(\mathbb{R}, h_{11}(t))$ and the semi-Riemannian manifold $(M, \varphi_{ij}(x))$. We point out that the affine maps between the manifolds $(\mathbb{R}, h_{11}(t))$ and $(M, \varphi_{ij}(x))$ are curves which carry the geodesics of the temporal manifold $(\mathbb{R}, h_{11}(t))$ into the geodesics on the spatial manifold $(M, \varphi_{ij}(x))$.

Remark 16. Multiplying the equations (1.13) with $h^{11} = 1/h_{11} \neq 0$, we obtain the equivalent equations

$$h^{11}\left[\frac{d^2 x^i}{dt^2} - \kappa_{11}^1(t)\frac{dx^i}{dt} + \gamma_{jk}^i(x)\frac{dx^j}{dt}\frac{dx^k}{dt} \right] = 0.$$

These are exactly the classical equations of the *harmonic maps* between the manifolds $(\mathbb{R}, h_{11}(t))$ and $(M, \varphi_{ij}(x))$ (see Eells and Lemaire [33]). For such a reason, we used the terminology of *harmonic curves* for the solutions of the SODEs (1.12).

Remark 17. The jet geometrical concept of *harmonic curve of a relativistic time-dependent semispray* $\mathcal{S} = (H, G)$ is intimately connected by the concept of *Euler-Lagrange equations* produced by a *relativistic time-dependent Lagrangian* $\mathcal{L} = L\sqrt{h_{11}(t)}$, where $L : J^1(\mathbb{R}, M) \to \mathbb{R}$. The connection is given by the fact that the Euler-Lagrange equations of any non-degenerate Lagrangian $\mathcal{L}$ can be written in the form (1.12). For example, the Euler-Lagrange equations of the jet Lagrangian

$$\mathcal{BS} = h^{11}(t)\varphi_{ij}(x)y_1^i y_1^j \sqrt{h_{11}(t)},$$

where $y_1^i = dx^i/dt$, are exactly the equations of the affine maps (1.13). Equations (1.13) are in fact the equations (1.12) for the particular relativistic time-dependent semispray $\overset{\circ}{\mathcal{S}} = (\overset{\circ}{H}, \overset{\circ}{G})$ associated to the pair of metrics $(h_{11}(t), \varphi_{ij}(x))$.

In this context, using the notations from Proposition 12, we immediately deduce the following interesting result:

Corollary 18. *The equations (1.12) of the harmonic curves of a relativistic time-dependent semispray* $\mathcal{S} = (H, G)$ *on the 1-jet space* $J^1(\mathbb{R}, M)$ *can be always rewritten in the following equivalent* **generalized Poisson form***:*

$$h^{11}\left[\frac{d^2 x^i}{dt^2} - \kappa_{11}^1(t)\frac{dx^i}{dt} + \gamma_{jk}^i(x)\frac{dx^j}{dt}\frac{dx^k}{dt}\right] = F^i\left(t, x^k(t), \frac{dx^k}{dt}\right),$$

where

$$F^i = 2h^{11}\left[T_{(1)1}^{(i)} + S_{(1)1}^{(i)}\right].$$

1.3   JET NONLINEAR CONNECTIONS. ADAPTED BASES

We have seen that the transformation rules of the canonical bases of vector fields (1.3) or covector fields (1.4) imply complicated transformation rules for the local components of the diverse geometrical objects (as the classical tensors, for example) on the 1-jet space $J^1(\mathbb{R}, M)$. For such a reason, it is necessary to construct the so-called *adapted bases* attached to a *nonlinear connection* on $J^1(\mathbb{R}, M)$. These adapted bases have the quality to simplify the transformation rules of the local components of the jet geometrical objects taken in the study.

In order to do this geometrical construction, let us consider an arbitrary point $u \in E = J^1(\mathbb{R}, M)$ and let us take the differential map

$$\pi_{*,u} : T_u E \to T_{(t,x)}(\mathbb{R} \times M)$$

produced by the canonical projection

$$\pi : E \to \mathbb{R} \times M, \ \pi(u) = (t, x).$$

The differential map $\pi_{*,u}$ generates the vector subspace

$$V_u = \ker \pi_{*,u} \subset T_u E,$$

whose dimension is $\dim_{\mathbb{R}} V_u = n$, $\forall\, u \in E$, because $\pi_{*,u}$ is a surjection. Moreover, a basis in the vector subspace V_u is given by

$$\left\{ \frac{\partial}{\partial y_1^i}\Big|_u \right\}.$$

It follows that the map

$$\mathcal{V} : u \in E \to V_u \subset T_u E$$

is a differential distribution on $J^1(\mathbb{R}, M)$, which is called the *vertical distribution* of the 1-jet space $E = J^1(\mathbb{R}, M)$.

Definition 19. A *nonlinear connection* on the 1-jet space $E = J^1(\mathbb{R}, M)$ is a differential distribution

$$\mathcal{H} : u \in E \to H_u \subset T_u E,$$

which verifies the equalities

$$T_u E = H_u \oplus V_u, \ \forall\, u \in E.$$

The differential distribution $\mathcal{H}$ is also called the *horizontal distribution* of the 1-jet space $J^1(\mathbb{R}, M)$.

Remark 20. (i) It is obvious that the dimension of a horizontal distribution is

$$\dim_{\mathbb{R}} H_u = n + 1, \ \forall\, u \in E.$$

(ii) The set $\mathcal{X}(E)$ of the vector fields on $E = J^1(\mathbb{R}, M)$ decomposes in the direct sum

$$\mathcal{X}(E) = \Gamma(\mathcal{H}) \oplus \mathcal{X}(\mathcal{V}), \tag{1.14}$$

where $\Gamma(\mathcal{H})$ [respectively $\mathcal{X}(\mathcal{V})$] represents the set of the *horizontal* [respectively *vertical*] *sections*.

Taking into account that a given nonlinear connection (horizontal distribution) $\mathcal{H}$ on the 1-jet space $E = J^1(\mathbb{R}, M)$ produces the isomorphisms

$$\pi_{*,u}\big|_{H_u} : H_u \to T_{\pi(u)}(\mathbb{R} \times M), \ \forall\, u \in E,$$

by direct local computations, we deduce the following geometrical results:

Proposition 21. (i) *There exist some unique linearly independent horizontal vector fields* $\delta/\delta t$, $\delta/\delta x^i \in \Gamma(\mathcal{H})$ *having the properties*

$$\pi_* \left(\frac{\delta}{\delta t} \right) = \frac{\partial}{\partial t}, \qquad \pi_* \left(\frac{\delta}{\delta x^i} \right) = \frac{\partial}{\partial x^i}. \tag{1.15}$$

(ii) *With respect to the natural basis $\{\partial/\partial t, \partial/\partial x^i, \partial/\partial y_1^i\} \subset \mathcal{X}(E)$, the horizontal vector fields $\delta/\delta t$ and $\delta/\delta x^i$ have the local expressions*

$$\frac{\delta}{\delta t} = \frac{\partial}{\partial t} - M_{(1)1}^{(j)}\frac{\partial}{\partial y_1^j}, \qquad \frac{\delta}{\delta x^i} = \frac{\partial}{\partial x^i} - N_{(1)i}^{(j)}\frac{\partial}{\partial y_1^j}, \qquad (1.16)$$

*where the functions $M_{(1)1}^{(j)}$ [respectively $N_{(1)i}^{(j)}$] are defined on the domains of the induced local charts on $E = J^1(\mathbb{R}, M)$ and they are called the **temporal** (respectively **spatial**) **components of the nonlinear connection** $\mathcal{H}$.*

(iii) *The local components $M_{(1)1}^{(j)}$ and $N_{(1)i}^{(j)}$ transform on every intersection of preceding induced local charts on E by the rules*

$$\widetilde{M}_{(1)1}^{(k)} = M_{(1)1}^{(j)}\left(\frac{dt}{d\widetilde{t}}\right)^2\frac{\partial\widetilde{x}^k}{\partial x^j} - \frac{dt}{d\widetilde{t}}\frac{\partial\widetilde{y}_1^k}{\partial t} \qquad (1.17)$$

and

$$\widetilde{N}_{(1)l}^{(k)} = N_{(1)i}^{(j)}\frac{dt}{d\widetilde{t}}\frac{\partial x^i}{\partial\widetilde{x}^l}\frac{\partial\widetilde{x}^k}{\partial x^j} - \frac{\partial x^i}{\partial\widetilde{x}^l}\frac{\partial\widetilde{y}_1^k}{\partial x^i}. \qquad (1.18)$$

(iv) *To give a nonlinear connection $\mathcal{H}$ on the 1-jet space $J^1(\mathbb{R}, M)$ is equivalent to giving on $E = J^1(\mathbb{R}, M)$ a set of local functions*

$$\Gamma = \left(M_{(1)1}^{(j)}, N_{(1)i}^{(j)}\right),$$

which transform by the rules (1.17) and (1.18).

■ **EXAMPLE 1.9**

Let $(\mathbb{R}, h_{11}(t))$ be a Riemannian manifold and let $(M, \varphi_{ij}(x))$ be a semi-Riemannian manifold. Let us consider the Christoffel symbols $\kappa_{11}^1(t)$ and $\gamma_{jk}^i(x)$. Then, using the transformation rules (1.2), (1.8), and (1.11), we deduce that the set of local functions

$$\overset{\circ}{\Gamma} = \left(\overset{\circ}{M}_{(1)1}^{(j)}, \overset{\circ}{N}_{(1)i}^{(j)}\right),$$

where

$$\overset{\circ}{M}_{(1)1}^{(j)} = -\kappa_{11}^1 y_1^j, \qquad \overset{\circ}{N}_{(1)i}^{(j)} = \gamma_{im}^j y_1^m, \qquad (1.19)$$

represents a nonlinear connection on the 1-jet space $J^1(\mathbb{R}, M)$. This jet nonlinear connection is called the *canonical nonlinear connection attached to the pair of metrics* $(h_{11}(t), \varphi_{ij}(x))$.

In the sequel, let us fix $\Gamma = \left(M_{(1)1}^{(j)}, N_{(1)i}^{(j)}\right)$, a nonlinear connection on the 1-jet space $E = J^1(\mathbb{R}, M)$. Then, the nonlinear connection Γ produces the horizontal vector fields (1.16) and the covector fields

$$\delta y_1^i = dy_1^i + M_{(1)1}^{(i)}dt + N_{(1)j}^{(i)}dx^j. \qquad (1.20)$$

It is easy to see now that the set of vector fields

$$\left\{ \frac{\delta}{\delta t}, \frac{\delta}{\delta x^i}, \frac{\partial}{\partial y_1^i} \right\} \subset \mathcal{X}(E) \tag{1.21}$$

represents a *basis* in the set of vector fields on $J^1(\mathbb{R}, M)$, and the set of covector fields

$$\left\{ dt, dx^i, \delta y_1^i \right\} \subset \mathcal{X}^*(E) \tag{1.22}$$

represents its *dual basis* in the set of 1-forms on $J^1(\mathbb{R}, M)$.

Definition 22. The dual bases (1.21) and (1.22) are called the *adapted bases* attached to the nonlinear connection Γ on the 1-jet space $E = J^1(\mathbb{R}, M)$.

The big advantage of the adapted bases produced by the nonlinear connection Γ is that the transformation laws of their elements are simple and natural.

Proposition 23. *The local transformation laws of the elements of the adapted bases (1.21) and (1.22) produced by the nonlinear connection* $\Gamma = \left(M_{(1)1}^{(j)}, N_{(1)i}^{(j)} \right)$ *are* **classical tensorial ones:**

$$\begin{cases} \dfrac{\delta}{\delta t} = \dfrac{d\widetilde{t}}{dt} \dfrac{\delta}{\delta \widetilde{t}}, \\[2mm] \dfrac{\delta}{\delta x^i} = \dfrac{\partial \widetilde{x}^j}{\partial x^i} \dfrac{\delta}{\delta \widetilde{x}^j}, \\[2mm] \dfrac{\partial}{\partial y_1^i} = \dfrac{\partial \widetilde{x}^j}{\partial x^i} \dfrac{dt}{d\widetilde{t}} \dfrac{\partial}{\partial \widetilde{y}_1^j}, \end{cases} \tag{1.23}$$

and

$$\begin{cases} dt = \dfrac{dt}{d\widetilde{t}} d\widetilde{t}, \\[2mm] dx^i = \dfrac{\partial x^i}{\partial \widetilde{x}^j} d\widetilde{x}^j, \\[2mm] \delta y_1^i = \dfrac{\partial x^i}{\partial \widetilde{x}^j} \dfrac{d\widetilde{t}}{dt} \delta \widetilde{y}_1^j. \end{cases} \tag{1.24}$$

Proof: Using the properties (1.15), we immediately deduce that we have

$$\pi_* \left(\frac{\delta}{\delta t} \right) = \frac{\partial}{\partial t} = \frac{d\widetilde{t}}{dt} \frac{\partial}{\partial \widetilde{t}} = \pi_* \left(\frac{d\widetilde{t}}{dt} \frac{\delta}{\delta \widetilde{t}} \right).$$

In other words, the temporal horizontal vector field

$$\frac{\delta}{\delta t} - \frac{d\widetilde{t}}{dt} \frac{\delta}{\delta \widetilde{t}} \in \Gamma(\mathcal{H}) \cap \mathcal{X}(\mathcal{V})$$

is also a vertical vector field. Taking into account the decomposition (1.14), it follows the required result.

By analogy, we treat the spatial horizontal vector fields $\delta/\delta x^i$.
Finally, let us remark that we have the equalities

$$
\begin{aligned}
\delta y_1^i &= \delta y_1^i \left(\frac{\delta}{\delta \tilde{t}} \right) d\tilde{t} + \delta y_1^i \left(\frac{\delta}{\delta \tilde{x}^j} \right) d\tilde{x}^j + \delta y_1^i \left(\frac{\partial}{\partial \tilde{y}_1^j} \right) \delta \tilde{y}_1^j \\
&= \delta y_1^i \left(\frac{dt}{d\tilde{t}} \frac{\delta}{\delta t} \right) d\tilde{t} + \delta y_1^i \left(\frac{\partial x^k}{\partial \tilde{x}^j} \frac{\delta}{\delta x^k} \right) d\tilde{x}^j + \delta y_1^i \left(\frac{\partial x^k}{\partial \tilde{x}^j} \frac{dt}{dt} \frac{\partial}{\partial y_1^k} \right) \delta \tilde{y}_1^j \\
&= \frac{\partial x^i}{\partial \tilde{x}^j} \frac{d\tilde{t}}{dt} \delta \tilde{y}_1^j.
\end{aligned}
$$

Corollary 24. *Any d-tensor field $D = \left(D_{1k(1)(l)...}^{1i(j)(1)...} \right)$ on the 1-jet space $J^1(\mathbb{R}, M)$ is a classical tensor field on $J^1(\mathbb{R}, M)$.*

Proof: Using the adapted bases attached to a nonlinear connection Γ and taking into account the transformation rules (1.5) of a d-tensor, it follows that a d-tensor $D = \left(D_{1k(1)(l)...}^{1i(j)(1)...} \right)$ can be regarded as a global geometrical object (a classical tensor) on the 1-jet space $J^1(\mathbb{R}, M)$, by putting

$$
D = D_{1k(1)(l)...}^{1i(j)(1)...} \frac{\delta}{\delta t} \otimes \frac{\delta}{\delta x^i} \otimes \frac{\partial}{\partial y_1^j} \otimes dt \otimes dx^k \otimes \delta y_1^l \otimes \dots .
$$

Remark 25. The utilization of parentheses for certain indices of the local components $D_{1k(1)(l)...}^{1i(j)(1)...}$ of the distinguished tensor D on $J^1(\mathbb{R}, M)$ is suitable for contractions. To illustrate this fact, we give the following examples:

(i) The *fundamental metrical d-tensor* produced by a *relativistic time-dependent Lagrangian function* (see Example 1.1) produces the geometrical object

$$
\mathbf{G} = G_{(i)(j)}^{(1)(1)} \delta y_1^i \otimes \delta y_1^j.
$$

(ii) The *canonical Liouville d-tensor field* of the 1-jet space $J^1(\mathbb{R}, M)$ (see Example 1.2) is represented by the geometrical object

$$
\mathbf{C} = \mathbf{C}_{(1)}^{(i)} \frac{\partial}{\partial y_1^i} = y_1^i \frac{\partial}{\partial y_1^i}.
$$

(iii) The *h-normalization d-tensor field* of the 1-jet space $J^1(\mathbb{R}, M)$ (see Example 1.3) has the representative object

$$
\mathbf{J}_h = J_{(1)1j}^{(i)} \frac{\partial}{\partial y_1^i} \otimes dt \otimes dx^j = h_{11} \frac{\partial}{\partial y_1^i} \otimes dt \otimes dx^i.
$$

(iv) The *h-canonical Liouville d-tensor field* of the 1-jet space $J^1(\mathbb{R}, M)$ (see Example 1.4) is equivalent to the geometrical object

$$
\mathbf{L}_h = L_{(1)11}^{(i)} \frac{\partial}{\partial y_1^i} \otimes dt \otimes dt = h_{11} y_1^i \frac{\partial}{\partial y_1^i} \otimes dt \otimes dt = \mathbf{C} \otimes h.
$$

1.4   RELATIVISTIC TIME-DEPENDENT SEMISPRAYS AND JET NONLINEAR CONNECTIONS

In this Section we study the geometrical relations between *relativistic time-dependent semisprays* and *nonlinear connections* on the 1-jet space $J^1(\mathbb{R}, M)$. In this direction, we prove the following geometrical results:

Proposition 26. (i) *The* **temporal semisprays** $H = \left(H^{(j)}_{(1)1} \right)$ *and the sets of* **temporal components of nonlinear connections** $M = \left(M^{(j)}_{(1)1} \right)$ *are in one-to-one correspondence on the 1-jet space* $J^1(\mathbb{R}, M)$, *via*

$$M^{(j)}_{(1)1} = 2H^{(j)}_{(1)1}, \qquad H^{(j)}_{(1)1} = \frac{1}{2}M^{(j)}_{(1)1}.$$

(ii) *The* **spatial semisprays** $G = \left(G^{(j)}_{(1)1} \right)$ *and the sets of* **spatial components of nonlinear connections** $N = \left(N^{(j)}_{(1)k} \right)$ *are connected on the 1-jet space* $J^1(\mathbb{R}, M)$, *via the relations*

$$N^{(j)}_{(1)k} = \frac{\partial G^{(j)}_{(1)1}}{\partial y^k_1}, \qquad G^{(j)}_{(1)1} = \frac{1}{2}N^{(j)}_{(1)m}y^m_1.$$

Proof: The result is an immediate consequence of the local transformation laws (1.7) and (1.17), respectively (1.10), (1.18), and (1.2).

Definition 27. The nonlinear connection $\Gamma_{\mathcal{S}}$ on the 1-jet space $J^1(\mathbb{R}, M)$, whose components are

$$\Gamma_{\mathcal{S}} = \left(M^{(j)}_{(1)1} = 2H^{(j)}_{(1)1}, \ N^{(j)}_{(1)k} = \frac{\partial G^{(j)}_{(1)1}}{\partial y^k_1} \right), \tag{1.25}$$

is called the *canonical jet nonlinear connection produced by the relativistic time-dependent semispray*

$$\mathcal{S} = (H, G) = \left(H^{(j)}_{(1)1}, G^{(j)}_{(1)1} \right).$$

Definition 28. The relativistic time-dependent semispray $\mathcal{S}_\Gamma$ on the 1-jet space $J^1(\mathbb{R}, M)$, whose components are

$$\mathcal{S}_\Gamma = \left(H^{(j)}_{(1)1} = \frac{1}{2}M^{(j)}_{(1)1}, \ G^{(j)}_{(1)1} = \frac{1}{2}N^{(j)}_{(1)m}y^m_1 \right), \tag{1.26}$$

is called the *canonical relativistic time-dependent semispray produced by the jet nonlinear connection*

$$\Gamma = \left(M^{(j)}_{(1)1}, N^{(j)}_{(1)k} \right).$$

Remark 29. The canonical jet nonlinear connection (1.25) produced by the relativistic time-dependent semispray S is a natural generalization of the canonical nonlinear connection N induced by a time-dependent semispray G in the *classical rheonomic (non-autonomous) Lagrangian geometry* [55].

Remark 30. Formulas (1.26) may offer us other interesting examples of jet relativistic time-dependent semisprays.

It is obvious that the equations (1.12) of the harmonic curves of the canonical relativistic time-dependent semispray (1.26) produced by the jet nonlinear connection

$$\Gamma = \left(M^{(j)}_{(1)1}, N^{(j)}_{(1)k} \right)$$

have the form

$$\frac{d^2 x^j}{dt^2} + M^{(j)}_{(1)1}\left(t, x^k(t), \frac{dx^k}{dt} \right) + N^{(j)}_{(1)m}\left(t, x^k(t), \frac{dx^k}{dt} \right) \frac{dx^m}{dt} = 0. \quad (1.27)$$

Definition 31. The smooth curves $c(t) = (x^i(t))$ which satisfy the equations (1.27) are called the *autoparallel harmonic curves of the jet nonlinear connection Γ*.

Remark 32. The geometrical concept of *autoparallel harmonic curve* of a jet nonlinear connection Γ naturally generalizes the concept of *path* of a time-dependent nonlinear connection N from the *classical non-autonomous (rheonomic) Lagrangian geometry* [55] or that of *autoparallel curve* of a nonlinear connection N from the *classical autonomous (time-independent) Lagrangian geometry* [24].

■ **EXAMPLE 1.10**

The autoparallel harmonic curves of the particular jet nonlinear connection (see Example 1.9)

$$\mathring{\Gamma} = \left(\mathring{M}^{(j)}_{(1)1}, \mathring{N}^{(j)}_{(1)i} \right)$$

attached to the pair of metrics $(h_{11}(t), \varphi_{ij}(x))$ are exactly the affine maps between the manifolds $(\mathbb{R}, h_{11}(t))$ and $(M, \varphi_{ij}(x))$.

CHAPTER 2

DEFLECTION d-TENSOR IDENTITIES IN THE RELATIVISTIC TIME-DEPENDENT LAGRANGE GEOMETRY

This Chapter provides an extension of the Miron-Anastasiei geometrical framework from [55] and describes on the 1-jet space $J^1(\mathbb{R}, M)$ the adapted components of the Γ-linear connections, together with their d-torsions and d-curvatures. For an arbitrary Γ-linear connection, the local Ricci identities, together with their corresponding nonmetrical deflection d-tensor identities, are also determined. We point out that the nonmetrical deflection d-tensor identities are necessary for the description of the geometrical Maxwell equations which govern our jet relativistic time-dependent electromagnetism.

2.1 THE ADAPTED COMPONENTS OF JET Γ-LINEAR CONNECTIONS

Let us suppose that on the 1-jet space $E = J^1(\mathbb{R}, M)$ is fixed a nonlinear connection

$$\Gamma = \left(M_{(1)1}^{(i)}, N_{(1)j}^{(i)} \right), \tag{2.1}$$

where $M_{(1)1}^{(i)}$ are its *temporal components* and $N_{(1)j}^{(i)}$ are its *spatial components*.

Jet Single-Time Lagrange Geometry and Its Applications   **19**
1-st Edition. By Vladimir Balan and Mircea Neagu.

Let

$$\left\{ \frac{\delta}{\delta t}, \frac{\delta}{\delta x^i}, \frac{\partial}{\partial y_1^i} \right\} \subset \mathcal{X}(E)$$

and

$$\{dt, dx^i, \delta y_1^i\} \subset \mathcal{X}^*(E)$$

be the dual bases adapted to the nonlinear connection (2.1), where

$$\frac{\delta}{\delta t} = \frac{\partial}{\partial t} - M_{(1)1}^{(j)} \frac{\partial}{\partial y_1^j},$$

$$\frac{\delta}{\delta x^i} = \frac{\partial}{\partial x^i} - N_{(1)i}^{(j)} \frac{\partial}{\partial y_1^j},$$

$$\delta y_1^i = dy_1^i + M_{(1)1}^{(i)} dt + N_{(1)j}^{(i)} dx^j.$$

In order to develop a theory of the Γ-linear connections on the 1-jet space $E = J^1(\mathbb{R}, M)$, we need the following simple result:

Proposition 33. (i) *The Lie algebra $\mathcal{X}(E)$ of the vector fields on $E = J^1(\mathbb{R}, M)$ decomposes in the direct sum*

$$\mathcal{X}(E) = \mathcal{X}(\mathcal{H}_{\mathbb{R}}) \oplus \mathcal{X}(\mathcal{H}_M) \oplus \mathcal{X}(\mathcal{V}),$$

where

$$\mathcal{X}(\mathcal{H}_{\mathbb{R}}) = Span\left\{ \frac{\delta}{\delta t} \right\}, \quad \mathcal{X}(\mathcal{H}_M) = Span\left\{ \frac{\delta}{\delta x^i} \right\}, \quad \mathcal{X}(\mathcal{V}) = Span\left\{ \frac{\partial}{\partial y_1^i} \right\}.$$

(ii) *The Lie algebra $\mathcal{X}^*(E)$ of the covector fields on $E = J^1(\mathbb{R}, M)$ decomposes in the direct sum*

$$\mathcal{X}^*(E) = \mathcal{X}^*(\mathcal{H}_{\mathbb{R}}) \oplus \mathcal{X}^*(\mathcal{H}_M) \oplus \mathcal{X}^*(\mathcal{V}),$$

where

$$\mathcal{X}^*(\mathcal{H}_{\mathbb{R}}) = Span\{dt\}, \quad \mathcal{X}^*(\mathcal{H}_M) = Span\{dx^i\}, \quad \mathcal{X}^*(\mathcal{V}) = Span\{\delta y_1^i\}.$$

Denoting by $h_{\mathbb{R}}$, h_M, respectively v, the $\mathbb{R}$-*horizontal*, M-*horizontal*, respectively *vertical canonical projections* associated to the above decompositions, we get the following:

Corollary 34. (i) *Any vector field on $E = J^1(\mathbb{R}, M)$ can be uniquely written in the form*

$$X = h_{\mathbb{R}} X + h_M X + vX, \quad \forall\, X \in \mathcal{X}(E).$$

(ii) *Any 1-form on $E = J^1(\mathbb{R}, M)$ can be uniquely written in the form*

$$\omega = h_{\mathbb{R}} \omega + h_M \omega + v\omega, \quad \forall\, \omega \in \mathcal{X}^*(E).$$

Definition 35. A linear connection $\nabla : \mathcal{X}(E) \times \mathcal{X}(E) \to \mathcal{X}(E)$, which verifies the Ehresmann-Koszul axioms

$$\nabla h_{\mathbb{R}} = 0, \quad \nabla h_M = 0, \quad \nabla v = 0,$$

is called a Γ-*linear connection* on the 1-jet vector bundle $E = J^1(\mathbb{R}, M)$.

Using the adapted basis of vector fields on $E = J^1(\mathbb{R}, M)$ and the definition of a Γ-linear connection, we prove without difficulties the following:

Proposition 36. *A* Γ-*linear connection* ∇ *on* $E = J^1(\mathbb{R}, M)$ *is determined by* **nine** *local adapted components*

$$\nabla\Gamma = \left(\bar{G}^1_{11}, \; G^k_{i1}, \; G^{(i)(1)}_{(1)(j)1}, \; \bar{L}^1_{1j}, \; L^k_{ij}, \; L^{(i)(1)}_{(1)(j)k}, \; \bar{C}^{1(1)}_{1(k)}, \; C^{j(1)}_{i(k)}, \; C^{(i)(1)(1)}_{(1)(j)(k)} \right),$$

which are uniquely defined by the relations

$$(h_{\mathbb{R}}) \quad \nabla_{\frac{\delta}{\delta t}} \frac{\delta}{\delta t} = \bar{G}^1_{11} \frac{\delta}{\delta t}, \quad \nabla_{\frac{\delta}{\delta t}} \frac{\delta}{\delta x^i} = G^k_{i1} \frac{\delta}{\delta x^k}, \quad \nabla_{\frac{\delta}{\delta t}} \frac{\partial}{\partial y^i_1} = G^{(k)(1)}_{(1)(i)1} \frac{\partial}{\partial y^k_1},$$

$$(h_M) \quad \nabla_{\frac{\delta}{\delta x^j}} \frac{\delta}{\delta t} = \bar{L}^1_{1j} \frac{\delta}{\delta t}, \quad \nabla_{\frac{\delta}{\delta x^j}} \frac{\delta}{\delta x^i} = L^k_{ij} \frac{\delta}{\delta x^k}, \quad \nabla_{\frac{\delta}{\delta x^j}} \frac{\partial}{\partial y^i_1} = L^{(k)(1)}_{(1)(i)j} \frac{\partial}{\partial y^k_1},$$

$$(v) \quad \nabla_{\frac{\partial}{\partial y^j_1}} \frac{\delta}{\delta t} = \bar{C}^{1(1)}_{1(j)} \frac{\delta}{\delta t}, \quad \nabla_{\frac{\partial}{\partial y^j_1}} \frac{\delta}{\delta x^i} = C^{k(1)}_{i(j)} \frac{\delta}{\delta x^k}, \quad \nabla_{\frac{\partial}{\partial y^j_1}} \frac{\partial}{\partial y^i_1} = C^{(k)(1)(1)}_{(1)(i)(j)} \frac{\partial}{\partial y^k_1}.$$

Taking into account the tensorial transformation laws of the adapted basis of vector fields on $E = J^1(\mathbb{R}, M)$, by laborious local computations, we deduce the following:

Theorem 37. (i) *Under a change of coordinates (1.2) on the 1-jet space* $E = J^1(\mathbb{R}, M)$, *the adapted coefficients of the* Γ-*linear connection* ∇ *modify by the rules*

$$(h_{\mathbb{R}}) \quad \begin{cases} \bar{G}^1_{11} = \tilde{\bar{G}}^1_{11} \dfrac{d\tilde{t}}{dt} + \dfrac{dt}{d\tilde{t}} \dfrac{d^2\tilde{t}}{dt^2}, \\[2ex] G^k_{i1} = \tilde{G}^r_{j1} \dfrac{\partial x^k}{\partial \tilde{x}^r} \dfrac{\partial \tilde{x}^j}{\partial x^i} \dfrac{d\tilde{t}}{dt}, \\[2ex] G^{(k)(1)}_{(1)(i)1} = \tilde{G}^{(p)(1)}_{(1)(j)1} \dfrac{\partial x^k}{\partial \tilde{x}^p} \dfrac{\partial \tilde{x}^j}{\partial x^i} \dfrac{d\tilde{t}}{dt} + \delta^k_i \left(\dfrac{d\tilde{t}}{dt} \right)^2 \dfrac{d^2 t}{d\tilde{t}^2}, \end{cases}$$

$$(h_M) \quad \begin{cases} \bar{L}^1_{1j} = \tilde{\bar{L}}^1_{1l} \dfrac{\partial \tilde{x}^l}{\partial x^j}, \\[2ex] L^r_{ij} = \tilde{L}^s_{pq} \dfrac{\partial x^r}{\partial \tilde{x}^s} \dfrac{\partial \tilde{x}^p}{\partial x^i} \dfrac{\partial \tilde{x}^q}{\partial x^j} + \dfrac{\partial x^r}{\partial \tilde{x}^s} \dfrac{\partial^2 \tilde{x}^s}{\partial x^i \partial x^j}, \\[2ex] L^{(k)(1)}_{(1)(i)j} = \tilde{L}^{(r)(1)}_{(1)(p)s} \dfrac{\partial x^k}{\partial \tilde{x}^r} \dfrac{\partial \tilde{x}^p}{\partial x^i} \dfrac{\partial \tilde{x}^s}{\partial x^j} + \dfrac{\partial x^k}{\partial \tilde{x}^r} \dfrac{\partial^2 \tilde{x}^r}{\partial x^i \partial x^j}, \end{cases}$$

$$(v) \quad \begin{cases} \bar{C}^{1(1)}_{1(i)} = \tilde{C}^{1(1)}_{1(j)} \dfrac{\partial \tilde{x}^j}{\partial x^i} \dfrac{dt}{d\tilde{t}}, \\[2mm] C^{k(1)}_{i(j)} = \tilde{C}^{s(1)}_{p(r)} \dfrac{\partial x^k}{\partial \tilde{x}^s} \dfrac{\partial \tilde{x}^p}{\partial x^i} \dfrac{\partial \tilde{x}^r}{\partial x^j} \dfrac{dt}{d\tilde{t}}, \\[2mm] C^{(k)(1)(1)}_{(1)(i)(j)} = \tilde{C}^{(r)(1)(1)}_{(1)(p)(q)} \dfrac{\partial x^k}{\partial \tilde{x}^r} \dfrac{\partial \tilde{x}^p}{\partial x^i} \dfrac{\partial \tilde{x}^q}{\partial x^j} \dfrac{dt}{d\tilde{t}}. \end{cases}$$

(ii) *Conversely, to give a Γ-linear connection ∇ on the 1-jet vector bundle $E = J^1(\mathbb{R}, M)$ is equivalent to giving a set of* **nine** *adapted local components $\nabla\Gamma$, which transform by the rules described in* (i).

■ **EXAMPLE 2.1**

Let $h_{11}(t)$ be a Riemannian metric on $\mathbb{R}$ and let $\varphi_{ij}(x)$ be a semi-Riemannian on M. We denote by $\kappa^1_{11}(t)$ [respectively $\gamma^k_{ij}(x)$] the Christoffel symbols of the metrics $h_{11}(t)$ and $\varphi_{ij}(x)$. Let us consider on the 1-jet space $E = J^1(\mathbb{R}, M)$ the canonical nonlinear connection $\overset{\circ}{\Gamma}$ associated to the pair of metrics $(h_{11}(t), \varphi_{ij}(x))$, which is defined by the local coefficients (1.19). In this context, using the transformation laws (1.8) and (1.11), we deduce that the set of adapted local coefficients

$$B\overset{\circ}{\Gamma} = \left(\bar{G}^1_{11}, \; 0, \; G^{(k)(1)}_{(1)(i)1}, \; 0, \; L^k_{ij}, \; L^{(k)(1)}_{(1)(i)j}, \; 0, \; 0, \; 0 \right),$$

where

$$\bar{G}^1_{11} = \kappa^1_{11}, \quad G^{(k)(1)}_{(1)(i)1} = -\delta^k_i \kappa^1_{11}, \quad L^k_{ij} = \gamma^k_{ij}, \quad L^{(k)(1)}_{(1)(i)j} = \gamma^k_{ij},$$

defines a $\overset{\circ}{\Gamma}$-linear connection on the 1-jet space E. This is called the *Berwald connection attached to the metrics $h_{11}(t)$ and $\varphi_{ij}(x)$.*

Remark 38. In the particular case $(\mathbb{R}, h) = (\mathbb{R}, \delta)$ our Berwald linear connection naturally generalizes the canonical N-linear connection induced by the canonical spray $2G^i = \gamma^i_{jk} y^j y^k$ from the classical theory of Finsler spaces (see Bao, Chern, and Shen [20] or Miron, Anastasiei, and Bucătaru [24], [55]).

Now, let us consider that ∇ is a fixed Γ-linear connection on the 1-jet space $E = J^1(\mathbb{R}, M)$, which is defined by the adapted local coefficients

$$\nabla\Gamma = \left(\bar{G}^1_{11}, \; G^k_{i1}, \; G^{(i)(1)}_{(1)(j)1}, \; \bar{L}^1_{1j}, \; L^k_{ij}, \; L^{(i)(1)}_{(1)(j)k}, \; \bar{C}^{1(1)}_{1(k)}, \; C^{j(1)}_{i(k)}, \; C^{(i)(1)(1)}_{(1)(j)(k)} \right). \quad (2.2)$$

Then, the Γ-linear connection ∇ naturally induces a linear connection on the set of the d-tensors of the 1-jet vector bundle E, in the following way: Starting with a vector field $X \in \mathcal{X}(E)$ and a d-tensor field D on $E = J^1(\mathbb{R}, M)$, locally expressed by

$$X = X^1 \frac{\delta}{\delta t} + X^r \frac{\delta}{\delta x^r} + X^{(r)}_{(1)} \frac{\partial}{\partial y^r_1},$$

$$D = D^{1i(j)(1)\cdots}_{1k(1)(l)\cdots} \frac{\delta}{\delta t} \otimes \frac{\delta}{\delta x^i} \otimes \frac{\partial}{\partial y^j_1} \otimes dt \otimes dx^k \otimes \delta y^l_1,$$

we introduce the covariant derivative

$$
\begin{aligned}
\nabla_X D &= X^1 \nabla_{\underbrace{\frac{\delta}{\delta t}}} D + X^p \nabla_{\underbrace{\frac{\delta}{\delta x^p}}} D + X^{(p)}_{(1)} \nabla_{\underbrace{\frac{\partial}{\partial y_1^p}}} D \\
&= \left\{ X^1 D^{1i(j)(1)\cdots}_{1k(1)(l)\cdots/1} + X^p \cdot D^{1i(j)(1)\cdots}_{1k(1)(l)\cdots|p} + X^{(p)}_{(1)} D^{1i(j)(1)\cdots}_{1k(1)(l)\cdots}\big|^{(1)}_{(p)} \right\} \\
&\quad \cdot \frac{\delta}{\delta t} \otimes \frac{\delta}{\delta x^i} \otimes \frac{\partial}{\partial y_1^j} \otimes dt \otimes dx^k \otimes \delta y_1^l,
\end{aligned}
$$

where

$(h_{\mathbb{R}})$
$$
\begin{cases}
D^{1i(j)(1)\cdots}_{1k(1)(l)\cdots/1} = \dfrac{\delta D^{1i(j)(1)\cdots}_{1k(1)(l)\cdots}}{\delta t} + D^{1i(j)(1)\cdots}_{1k(1)(l)\cdots} \bar{G}^1_{11} \\[2mm]
+ D^{1r(j)(1)\cdots}_{1k(1)(l)\cdots} G^i_{r1} + D^{1i(r)(1)\cdots}_{1k(1)(l)\cdots} G^{(j)(1)}_{(1)(r)1} + \cdots \\[2mm]
- D^{1i(j)(1)\cdots}_{1k(1)(l)\cdots} \bar{G}^1_{11} - D^{1i(j)(1)\cdots}_{1r(1)(l)\cdots} G^r_{k1} - D^{1i(j)(1)\cdots}_{1k(1)(r)\cdots} G^{(r)(1)}_{(1)(l)1} - \cdots,
\end{cases}
$$

(h_M)
$$
\begin{cases}
D^{1i(j)(1)\cdots}_{1k(1)(l)\cdots|p} = \dfrac{\delta D^{1i(j)(1)\cdots}_{1k(1)(l)\cdots}}{\delta x^p} + D^{1i(j)(1)\cdots}_{1k(1)(l)\cdots} \bar{L}^1_{1p} \\[2mm]
+ D^{1r(j)(1)\cdots}_{1k(1)(l)\cdots} L^i_{rp} + D^{1i(r)(1)\cdots}_{1k(1)(l)\cdots} L^{(j)(1)}_{(1)(r)p} + \cdots \\[2mm]
- D^{1i(j)(1)\cdots}_{1k(1)(l)\cdots} \bar{L}^1_{1p} - D^{1i(j)(1)\cdots}_{1r(1)(l)\cdots} L^r_{kp} - D^{1i(j)(1)\cdots}_{1k(1)(r)\cdots} L^{(r)(1)}_{(1)(l)p} - \cdots,
\end{cases}
$$

(v)
$$
\begin{cases}
D^{1i(j)(1)\cdots}_{1k(1)(l)\cdots}\big|^{(1)}_{(p)} = \dfrac{\partial D^{1i(j)(1)\cdots}_{1k(1)(l)\cdots}}{\partial y_1^p} + D^{1i(j)(1)\cdots}_{1k(1)(l)\cdots} \bar{C}^{1(1)}_{1(p)} \\[2mm]
+ D^{1r(j)(1)\cdots}_{1k(1)(l)\cdots} C^{i(1)}_{r(p)} + D^{1i(r)(1)\cdots}_{1k(1)(l)\cdots} C^{(j)(1)(1)}_{(1)(r)(p)} + \cdots \\[2mm]
- D^{1i(j)(1)\cdots}_{1k(1)(l)\cdots} \bar{C}^{1(1)}_{1(p)} - D^{1i(j)(1)\cdots}_{1r(1)(l)\cdots} C^{r(1)}_{k(p)} - D^{1i(j)(1)\cdots}_{1k(1)(r)\cdots} C^{(r)(1)(1)}_{(1)(l)(p)} - \cdots.
\end{cases}
$$

Definition 39. The local derivative operators " $_{/1}$," " $_{|p}$," and " $\big|^{(1)}_{(p)}$ " are called the $\mathbb{R}$-*horizontal covariant derivative*, the M-*horizontal covariant derivative*, and the *vertical covariant derivative associated to the Γ-linear connection* $\nabla\Gamma$. These apply to the local components of an arbitrary d-tensor field on the space $E = J^1(\mathbb{R}, M)$.

Remark 40. (i) In the particular case of a function $f(t, x^k, y_1^k)$ on the 1-jet space $J^1(\mathbb{R}, M)$ the above covariant derivatives reduce to

$$
f_{/1} = \frac{\delta f}{\delta t} = \frac{\partial f}{\partial t} - M^{(k)}_{(1)1} \frac{\partial f}{\partial y_1^k}, \quad
f_{|p} = \frac{\delta f}{\delta x^p} = \frac{\partial f}{\partial x^p} - N^{(k)}_{(1)p} \frac{\partial f}{\partial y_1^k}, \quad
f\big|^{(1)}_{(p)} = \frac{\partial f}{\partial y_1^p}.
$$

(ii) Starting with a d-vector field $D = Y$ on the 1-jet space $E = J^1(\mathbb{R}, M)$, locally expressed by

$$
Y = Y^1 \frac{\delta}{\delta t} + Y^i \frac{\delta}{\delta x^i} + Y^{(i)}_{(1)} \frac{\partial}{\partial y_1^i},
$$

the following expressions of the local covariant derivatives hold true:

$$(h_{\mathbb{R}}) \quad \begin{cases} Y^1_{/1} = \dfrac{\delta Y^1}{\delta t} + Y^1 \bar{G}^1_{11}, \\[2ex] Y^i_{/1} = \dfrac{\delta Y^i}{\delta t} + Y^r G^i_{r1}, \\[2ex] Y^{(i)}_{(1)/1} = \dfrac{\delta Y^{(i)}_{(1)}}{\delta t} + Y^{(r)}_{(1)} G^{(i)(1)}_{(1)(r)1}, \end{cases}$$

$$(h_M) \quad \begin{cases} Y^1_{|p} = \dfrac{\delta Y^1}{\delta x^p} + Y^1 \bar{L}^1_{1p}, \\[2ex] Y^i_{|p} = \dfrac{\delta Y^i}{\delta x^p} + Y^r L^i_{rp}, \\[2ex] Y^{(i)}_{(1)|p} = \dfrac{\delta Y^{(i)}_{(1)}}{\delta x^p} + Y^{(r)}_{(1)} L^{(i)(1)}_{(1)(r)p}, \end{cases}$$

$$(v) \quad \begin{cases} Y^1|^{(1)}_{(p)} = \dfrac{\partial Y^1}{\partial y^p_1} + Y^1 \bar{C}^{1(1)}_{1(p)}, \\[2ex] Y^i|^{(1)}_{(p)} = \dfrac{\partial Y^i}{\partial y^p_1} + Y^r C^{i(1)}_{r(p)}, \\[2ex] Y^{(i)}_{(1)}|^{(1)}_{(p)} = \dfrac{\partial Y^{(i)}_{(1)}}{\partial y^p_1} + Y^{(r)}_{(1)} C^{(i)(1)(1)}_{(1)(r)(p)}. \end{cases}$$

Denoting generically by " $_{:A}$ " one of the local covariant derivatives " $_{/1}$," " $_{|p}$," or " $|^{(1)}_{(p)}$," we obtain the following properties of the covariant derivative operators:

Proposition 41. *If $T^{\cdots}_{\cdots}$ and $S^{\cdots}_{\cdots}$ are two arbitrary d-tensors on $E = J^1(\mathbb{R}, M)$, then the following statements hold true:*

(i) *The local coefficients $T^{\cdots}_{\cdots:A}$ represent the components of a new d-tensor field on $J^1(\mathbb{R}, M)$.*

(ii) $(T^{\cdots}_{\cdots} + S^{\cdots}_{\cdots})_{:A} = T^{\cdots}_{\cdots:A} + S^{\cdots}_{\cdots:A}.$

(iii) $(T^{\cdots}_{\cdots} \otimes S^{\cdots}_{\cdots})_{:A} = T^{\cdots}_{\cdots:A} \otimes S^{\cdots}_{\cdots} + T^{\cdots}_{\cdots} \otimes S^{\cdots}_{\cdots:A}.$

2.2 LOCAL TORSION AND CURVATURE d-TENSORS

In the sequel, we will study the torsion tensor $\mathbf{T} : \mathcal{X}(E) \times \mathcal{X}(E) \to \mathcal{X}(E)$ associated to the Γ-linear connection ∇, which is given by the formula

$$\mathbf{T}(X, Y) = \nabla_X Y - \nabla_Y X - [X, Y], \quad \forall\, X, Y \in \mathcal{X}(E).$$

In order to obtain an adapted local characterization of the torsion tensor $\mathbf{T}$ of the Γ-linear connection ∇, we first deduce, by direct computations, the following important result:

Proposition 42. *The following identities of the* **Poisson brackets** *are true:*

$$\left[\frac{\delta}{\delta t}, \frac{\delta}{\delta t}\right] = 0, \qquad\qquad \left[\frac{\delta}{\delta t}, \frac{\delta}{\delta x^j}\right] = R^{(r)}_{(1)1j}\frac{\partial}{\partial y_1^r},$$

$$\left[\frac{\delta}{\delta t}, \frac{\partial}{\partial y_1^j}\right] = \frac{\partial M^{(r)}_{(1)1}}{\partial y_1^j}\frac{\partial}{\partial y_1^r}, \qquad \left[\frac{\delta}{\delta x^i}, \frac{\delta}{\delta x^j}\right] = R^{(r)}_{(1)ij}\frac{\partial}{\partial y_1^r},$$

$$\left[\frac{\delta}{\delta x^i}, \frac{\partial}{\partial y_1^j}\right] = \frac{\partial N^{(r)}_{(1)i}}{\partial y_1^j}\frac{\partial}{\partial y_1^r}, \qquad \left[\frac{\partial}{\partial y_1^i}, \frac{\partial}{\partial y_1^j}\right] = 0,$$

where $M^{(r)}_{(1)1}$ *and* $N^{(r)}_{(1)i}$ *are the local coefficients of the nonlinear connection* Γ*, while the components* $R^{(r)}_{(1)1j}$ *and* $R^{(r)}_{(1)ij}$ *are d-tensors given by the formulas*

$$
\begin{aligned}
R^{(r)}_{(1)1j} &= \frac{\delta M^{(r)}_{(1)1}}{\delta x^j} - \frac{\delta N^{(r)}_{(1)j}}{\delta t}, \\
R^{(r)}_{(1)ij} &= \frac{\delta N^{(r)}_{(1)i}}{\delta x^j} - \frac{\delta N^{(r)}_{(1)j}}{\delta x^i}.
\end{aligned}
\tag{2.3}
$$

In these conditions, working with a basis of vector fields, adapted to the nonlinear connection (2.1), by local computations, we obtain

Theorem 43. *The torsion tensor* **T** *of the* Γ-*linear connection (2.2) is determined by the following* **ten** *adapted torsion d-tensors:*

$$h_{\mathbb{R}}\mathbf{T}\left(\frac{\delta}{\delta t}, \frac{\delta}{\delta t}\right) = 0, \quad h_M\mathbf{T}\left(\frac{\delta}{\delta t}, \frac{\delta}{\delta t}\right) = 0, \quad v\mathbf{T}\left(\frac{\delta}{\delta t}, \frac{\delta}{\delta t}\right) = 0,$$

$$h_{\mathbb{R}}\mathbf{T}\left(\frac{\delta}{\delta x^j}, \frac{\delta}{\delta t}\right) = \bar{T}^1_{1j}\frac{\delta}{\delta t}, \quad h_M\mathbf{T}\left(\frac{\delta}{\delta x^j}, \frac{\delta}{\delta t}\right) = T^r_{1j}\frac{\delta}{\delta x^r},$$

$$v\mathbf{T}\left(\frac{\delta}{\delta x^j}, \frac{\delta}{\delta t}\right) = R^{(r)}_{(1)1j}\frac{\partial}{\partial y_1^r},$$

$$h_{\mathbb{R}}\mathbf{T}\left(\frac{\delta}{\delta x^j}, \frac{\delta}{\delta x^i}\right) = 0, \quad h_M\mathbf{T}\left(\frac{\delta}{\delta x^j}, \frac{\delta}{\delta x^i}\right) = T^r_{ij}\frac{\delta}{\delta x^r},$$

$$v\mathbf{T}\left(\frac{\delta}{\delta x^j}, \frac{\delta}{\delta x^i}\right) = R^{(r)}_{(1)ij}\frac{\partial}{\partial y_1^r},$$

$$h_{\mathbb{R}}\mathbf{T}\left(\frac{\partial}{\partial y_1^j}, \frac{\delta}{\delta t}\right) = \bar{P}^{1(1)}_{1(j)}\frac{\delta}{\delta t}, \quad h_M\mathbf{T}\left(\frac{\partial}{\partial y_1^j}, \frac{\delta}{\delta t}\right) = 0,$$

$$v\mathbf{T}\left(\frac{\partial}{\partial y_1^j}, \frac{\delta}{\delta t}\right) = P^{(r)\ (1)}_{(1)1(j)}\frac{\partial}{\partial y_1^r},$$

$$h_{\mathbb{R}}\mathbf{T}\left(\frac{\partial}{\partial y_1^j},\frac{\delta}{\delta x^i}\right)=0,\quad h_M\mathbf{T}\left(\frac{\partial}{\partial y_1^j},\frac{\delta}{\delta x^i}\right)=P_{i(j)}^{r(1)}\frac{\delta}{\delta x^r},$$

$$v\mathbf{T}\left(\frac{\partial}{\partial y_1^j},\frac{\delta}{\delta x^i}\right)=P_{(1)i(j)}^{(r)\,(1)}\frac{\partial}{\partial y_1^r},$$

$$h_{\mathbb{R}}\mathbf{T}\left(\frac{\partial}{\partial y_1^j},\frac{\partial}{\partial y_1^i}\right)=0,\quad h_M\mathbf{T}\left(\frac{\partial}{\partial y_1^j},\frac{\partial}{\partial y_1^i}\right)=0,$$

$$v\mathbf{T}\left(\frac{\partial}{\partial y_1^j},\frac{\partial}{\partial y_1^i}\right)=S_{(1)(i)(j)}^{(r)(1)(1)}\frac{\partial}{\partial y_1^r},$$

where

$$\bar{T}_{1j}^1=\bar{L}_{1j}^1,\quad T_{1j}^r=-G_{j1}^r,\quad T_{ij}^r=L_{ij}^r-L_{ji}^r,\quad \bar{P}_{1(j)}^{1(1)}=\bar{C}_{1(j)}^{1(1)},$$

$$P_{i(j)}^{r(1)}=C_{i(j)}^{r(1)},\quad S_{(1)(i)(j)}^{(r)(1)(1)}=C_{(1)(i)(j)}^{(r)(1)(1)}-C_{(1)(j)(i)}^{(r)(1)(1)},$$

$$P_{(1)1(j)}^{(r)\,(1)}=\frac{\partial M_{(1)1}^{(r)}}{\partial y_1^j}-G_{(1)(j)1}^{(r)(1)},\quad P_{(1)i(j)}^{(r)\,(1)}=\frac{\partial N_{(1)i}^{(r)}}{\partial y_1^j}-L_{(1)(j)i}^{(r)(1)},$$

and the d-tensors $R_{(1)1j}^{(r)}$ and $R_{(1)ij}^{(r)}$ are given by (2.3).

Corollary 44. *The torsion tensor* $\mathbf{T}$ *of an arbitrary* Γ-*linear connection* ∇ *on the 1-jet space* $E=J^1(\mathbb{R},M)$ *is determined by* **ten** *effective adapted local torsion d-tensors, which we arrange in the following table:*

	$h_{\mathbb{R}}$	h_M	v
$h_{\mathbb{R}}h_{\mathbb{R}}$	0	0	0
$h_M h_{\mathbb{R}}$	$\bar{T}_{1j}^1$	T_{1j}^r	$R_{(1)1j}^{(r)}$
$h_M h_M$	0	T_{ij}^r	$R_{(1)ij}^{(r)}$
$vh_{\mathbb{R}}$	$\bar{P}_{1(j)}^{1(1)}$	0	$P_{(1)1(j)}^{(r)\,(1)}$
vh_M	0	$P_{i(j)}^{r(1)}$	$P_{(1)i(j)}^{(r)\,(1)}$
vv	0	0	$S_{(1)(i)(j)}^{(r)(1)(1)}$

■ **EXAMPLE 2.2**

In the particular case of the Berwald $\overset{\circ}{\Gamma}$-linear connection $B\overset{\circ}{\Gamma}$, associated to the metrics $h_{11}(t)$ and $\varphi_{ij}(x)$, all torsion d-tensors vanish, except

$$R_{(1)ij}^{(k)} = \mathfrak{R}_{mij}^{k} y_{1}^{m},$$

where $\mathfrak{R}_{mij}^{k}(x)$ are the classical local curvature tensors of the spatial semi-Riemannian metric $\varphi_{ij}(x)$.

In order to study the curvature of the Γ-linear connection ∇, we recall that the curvature tensor $\mathbf{R}$ of ∇ is given by the formula

$$\mathbf{R}(X, Y)Z = \nabla_X \nabla_Y Z - \nabla_Y \nabla_X Z - \nabla_{[X,Y]} Z, \quad \forall\, X, Y, Z \in \mathcal{X}(E).$$

Using again a basis of vector fields adapted to the nonlinear connection Γ, together with the properties of the Γ-linear connection ∇, by direct computations, we obtain the following:

Theorem 45. *The curvature tensor $\mathbf{R}$ associated to the Γ-linear connection (2.2) is determined by* **fifteen** *effective adapted local curvature d-tensors:*

$$\mathbf{R}\left(\frac{\delta}{\delta t}, \frac{\delta}{\delta t}\right) \frac{\delta}{\delta t} = 0, \quad \mathbf{R}\left(\frac{\delta}{\delta t}, \frac{\delta}{\delta t}\right) \frac{\delta}{\delta x^i} = 0, \quad \mathbf{R}\left(\frac{\delta}{\delta t}, \frac{\delta}{\delta t}\right) \frac{\partial}{\partial y_1^i} = 0,$$

$$\mathbf{R}\left(\frac{\delta}{\delta x^k}, \frac{\delta}{\delta t}\right) \frac{\delta}{\delta t} = \bar{R}_{11k}^{1} \frac{\delta}{\delta t}, \quad \mathbf{R}\left(\frac{\delta}{\delta x^k}, \frac{\delta}{\delta t}\right) \frac{\delta}{\delta x^i} = R_{i1k}^{l} \frac{\delta}{\delta x^l},$$

$$\mathbf{R}\left(\frac{\delta}{\delta x^k}, \frac{\delta}{\delta t}\right) \frac{\partial}{\partial y_1^i} = R_{(1)(i)1k}^{(l)(1)} \frac{\partial}{\partial y_1^i},$$

$$\mathbf{R}\left(\frac{\delta}{\delta x^k}, \frac{\delta}{\delta x^j}\right) \frac{\delta}{\delta t} = \bar{R}_{1jk}^{1} \frac{\delta}{\delta t}, \quad \mathbf{R}\left(\frac{\delta}{\delta x^k}, \frac{\delta}{\delta x^j}\right) \frac{\delta}{\delta x^i} = R_{ijk}^{l} \frac{\delta}{\delta x^l},$$

$$\mathbf{R}\left(\frac{\delta}{\delta x^k}, \frac{\delta}{\delta x^j}\right) \frac{\partial}{\partial y_1^i} = R_{(1)(i)jk}^{(l)(1)} \frac{\partial}{\partial y_1^l},$$

$$\mathbf{R}\left(\frac{\partial}{\partial y_1^k}, \frac{\delta}{\delta t}\right) \frac{\delta}{\delta t} = \bar{P}_{11(k)}^{1\ (1)} \frac{\delta}{\delta t}, \quad \mathbf{R}\left(\frac{\partial}{\partial y_1^k}, \frac{\delta}{\delta t}\right) \frac{\delta}{\delta x^i} = P_{i1(k)}^{l\ (1)} \frac{\delta}{\delta x^l},$$

$$\mathbf{R}\left(\frac{\partial}{\partial y_1^k}, \frac{\delta}{\delta t}\right) \frac{\partial}{\partial y_1^i} = P_{(1)(i)1(k)}^{(l)(1)\ (1)} \frac{\partial}{\partial y_1^l},$$

$$\mathbf{R}\left(\frac{\partial}{\partial y_1^k}, \frac{\delta}{\delta x^j}\right) \frac{\delta}{\delta t} = \bar{P}_{1j(k)}^{1\ (1)} \frac{\delta}{\delta t}, \quad \mathbf{R}\left(\frac{\partial}{\partial y_1^k}, \frac{\delta}{\delta x^j}\right) \frac{\delta}{\delta x^i} = P_{ij(k)}^{l\ (1)} \frac{\delta}{\delta x^l},$$

$$\mathbf{R}\left(\frac{\partial}{\partial y_1^k}, \frac{\delta}{\delta x^j}\right) \frac{\partial}{\partial y_1^i} = P_{(1)(i)j(k)}^{(l)(1)\ (1)} \frac{\partial}{\partial y_1^l},$$

$$\mathbf{R}\left(\frac{\partial}{\partial y_1^k}, \frac{\partial}{\partial y_1^j}\right)\frac{\delta}{\delta t} = \bar{S}_{1(j)(k)}^{1(1)(1)}\frac{\delta}{\delta t}, \quad \mathbf{R}\left(\frac{\partial}{\partial y_1^k}, \frac{\partial}{\partial y_1^j}\right)\frac{\delta}{\delta x^i} = S_{i(j)(k)}^{l(1)(1)}\frac{\delta}{\delta x^l},$$

$$\mathbf{R}\left(\frac{\partial}{\partial y_1^k}, \frac{\partial}{\partial y_1^j}\right)\frac{\partial}{\partial y_1^i} = S_{(1)(i)(j)(k)}^{(l)(1)(1)(1)}\frac{\partial}{\partial y_1^l},$$

whose local components we arrange in the following table:

	$h_\mathbb{R}$	h_M	v
$h_\mathbb{R} h_\mathbb{R}$	0	0	0
$h_M h_\mathbb{R}$	$\bar{R}_{11k}^1$	R_{i1k}^l	$R_{(1)(i)1k}^{(l)(1)}$
$h_M h_M$	$\bar{R}_{1jk}^1$	R_{ijk}^l	$R_{(1)(i)jk}^{(l)(1)}$
$v h_\mathbb{R}$	$\bar{P}_{11(k)}^{1\ (1)}$	$P_{i1(k)}^{l\ (1)}$	$P_{(1)(i)1(k)}^{(l)(1)\ (1)}$
$v h_M$	$\bar{P}_{1j(k)}^{1\ (1)}$	$P_{ij(k)}^{l\ (1)}$	$P_{(1)(i)j(k)}^{(l)(1)\ (1)}$
$v v$	$\bar{S}_{1(j)(k)}^{1(1)(1)}$	$S_{i(j)(k)}^{l(1)(1)}$	$S_{(1)(i)(j)(k)}^{(l)(1)(1)(1)}$

Moreover, by laborious local computations, we deduce the following result:

Theorem 46. *The components of the preceding local curvature d-tensors are as follows:*

- $h_\mathbb{R}$-*components*

1. $\bar{R}_{11k}^1 = \dfrac{\delta \bar{G}_{11}^1}{\delta x^k} - \dfrac{\delta \bar{L}_{1k}^1}{\delta t} + \bar{C}_{1(r)}^{1(1)} R_{(1)1k}^{(r)}$

2. $\bar{R}_{1jk}^1 = \dfrac{\delta \bar{L}_{1j}^1}{\delta x^k} - \dfrac{\delta \bar{L}_{1k}^1}{\delta x^j} + \bar{C}_{1(r)}^{1(1)} R_{(1)jk}^{(r)}$

3. $\bar{P}_{11(k)}^{1\ (1)} = \dfrac{\partial \bar{G}_{11}^1}{\partial y_1^k} - \bar{C}_{1(k)/1}^{1(1)} + \bar{C}_{1(r)}^{1(1)} P_{(1)1(k)}^{(r)\ (1)}$

4. $\bar{P}_{1j(k)}^{1\ (1)} = \dfrac{\partial \bar{L}_{1j}^1}{\partial y_1^k} - \bar{C}_{1(k)|j}^{1(1)} + \bar{C}_{1(r)}^{1(1)} P_{(1)j(k)}^{(r)\ (1)}$

5. $\bar{S}_{1(j)(k)}^{1(1)(1)} = \dfrac{\partial \bar{C}_{1(j)}^{1(1)}}{\partial y_1^k} - \dfrac{\partial \bar{C}_{1(k)}^{1(1)}}{\partial y_1^j}$

- h_M-*components*

6. $R_{i1k}^l = \dfrac{\delta G_{i1}^l}{\delta x^k} - \dfrac{\delta L_{ik}^l}{\delta t} + G_{i1}^r L_{rk}^l - L_{ik}^r G_{r1}^l + C_{i(r)}^{l(1)} R_{(1)1k}^{(r)}$

7. $R^l_{ijk} = \dfrac{\delta L^l_{ij}}{\delta x^k} - \dfrac{\delta L^l_{ik}}{\delta x^j} + L^r_{ij} L^l_{rk} - L^r_{ik} L^l_{rj} + C^{l(1)}_{i(r)} R^{(r)}_{(1)jk}$

8. $P^{l\ \ (1)}_{i1(k)} = \dfrac{\partial G^l_{i1}}{\partial y^k_1} - C^{l(1)}_{i(k)/1} + C^{l(1)}_{i(r)} P^{(r)\ (1)}_{(1)1(k)}$

9. $P^{l\ \ (1)}_{ij(k)} = \dfrac{\partial L^l_{ij}}{\partial y^k_1} - C^{l(1)}_{i(k)|j} + C^{l(1)}_{i(r)} P^{(r)\ (1)}_{(1)j(k)}$

10. $S^{l(1)(1)}_{i(j)(k)} = \dfrac{\partial C^{l(1)}_{i(j)}}{\partial y^k_1} - \dfrac{\partial C^{l(1)}_{i(k)}}{\partial y^j_1} + C^{r(1)}_{i(j)} C^{l(1)}_{r(k)} - C^{r(1)}_{i(k)} C^{l(1)}_{r(j)}$

- v-*components*

11. $R^{(l)(1)}_{(1)(i)1k} = \dfrac{\delta G^{(l)(1)}_{(1)(i)1}}{\delta x^k} - \dfrac{\delta L^{(l)(1)}_{(1)(i)k}}{\delta t} + G^{(r)(1)}_{(1)(i)1} L^{(l)(1)}_{(1)(r)k}$

$$- L^{(r)(1)}_{(1)(i)k} G^{(l)(1)}_{(1)(r)1} + C^{(l)(1)(1)}_{(1)(i)(r)} R^{(r)}_{(1)1k}$$

12. $R^{(l)(1)}_{(1)(i)jk} = \dfrac{\delta L^{(l)(1)}_{(1)(i)j}}{\delta x^k} - \dfrac{\delta L^{(l)(1)}_{(1)(i)k}}{\delta x^j} + L^{(r)(1)}_{(1)(i)j} L^{(l)(1)}_{(1)(r)k}$

$$- L^{(r)(1)}_{(1)(i)k} L^{(l)(1)}_{(1)(r)j} + C^{(l)(1)(1)}_{(1)(i)(r)} R^{(r)}_{(1)jk}$$

13. $P^{(l)(1)\ (1)}_{(1)(i)1(k)} = \dfrac{\partial G^{(l)(1)}_{(1)(i)1}}{\partial y^k_1} - C^{(l)(1)(1)}_{(1)(i)(k)/1} + C^{(l)(1)(1)}_{(1)(i)(r)} P^{(r)\ (1)}_{(1)1(k)}$

14. $P^{(l)(1)\ (1)}_{(1)(i)j(k)} = \dfrac{\partial L^{(l)(1)}_{(1)(i)j}}{\partial y^k_1} - C^{(l)(1)(1)}_{(1)(i)(k)|j} + C^{(l)(1)(1)}_{(1)(i)(r)} P^{(r)\ (1)}_{(1)j(k)}$

15. $S^{(l)(1)(1)(1)}_{(1)(i)(j)(k)} = \dfrac{\partial C^{(l)(1)(1)}_{(1)(i)(j)}}{\partial y^k_1} - \dfrac{\partial C^{(l)(1)(1)}_{(1)(i)(k)}}{\partial y^j_1} + C^{(r)(1)(1)}_{(1)(i)(j)} C^{(l)(1)(1)}_{(1)(r)(k)}$

$$- C^{(r)(1)(1)}_{(1)(i)(k)} C^{(l)(1)(1)}_{(1)(r)(j)}.$$

■ **EXAMPLE 2.3**

In the case of the Berwald $\mathring{\Gamma}$-linear connection $B\mathring{\Gamma}$, associated to the pair of metrics $(h_{11}(t), \varphi_{ij}(x))$, all local curvature d-tensors vanish, except

$$R^l_{ijk} = \mathfrak{R}^l_{ijk}, \quad R^{(l)(1)}_{(1)(i)jk} = \mathfrak{R}^l_{ijk},$$

where $\mathfrak{R}^l_{ijk}(x)$ are the classical local curvature tensors of the spatial semi-Riemannian metric $\varphi_{ij}(x)$.

2.3   LOCAL RICCI IDENTITIES AND NONMETRICAL DEFLECTION d-TENSORS

Using the properties of a Γ-linear connection ∇ given by (2.2), together with the definitions of its torsion tensor $\mathbf{T}$ and its curvature tensor $\mathbf{R}$, we can prove the following important result which is used in the subsequent Lagrangian geometrical theory of the jet relativistic time-dependent electromagnetism, in order to describe its geometrical Maxwell equations.

Theorem 47. *If X is an arbitrary d-vector field on the 1-jet space $E = J^1(\mathbb{R}, M)$, locally expressed by $X = X^1 \dfrac{\delta}{\delta t} + X^i \dfrac{\delta}{\delta x^i} + X^{(i)}_{(1)} \dfrac{\partial}{\partial y^i_1}$, then the following **Ricci identities** of the Γ-linear connection ∇ are true:*

$$(h_{\mathbb{R}}) \begin{cases} X^1_{/1|k} - X^1_{|k/1} = X^1 \bar{R}^1_{11k} - X^1_{/1} \bar{T}^1_{1k} - X^1_{|r} T^r_{1k} - X^1|^{(1)}_{(r)} R^{(r)}_{(1)1k}, \\[6pt] X^1_{|j|k} - X^1_{|k|j} = X^1 \bar{R}^1_{1jk} - X^1_{|r} T^r_{jk} - X^1|^{(1)}_{(r)} R^{(r)}_{(1)jk}, \\[6pt] X^1_{/1}|^{(1)}_{(k)} - X^1|^{(1)}_{(k)/1} = X^1 \bar{P}^{1\ (1)}_{11(k)} - X^1_{/1} \bar{C}^{1(1)}_{1(k)} - X^1|^{(1)}_{(r)} P^{(r)\ (1)}_{(1)1(k)}, \\[6pt] X^1_{|j}|^{(1)}_{(k)} - X^1|^{(1)}_{(k)|j} = X^1 \bar{P}^{1\ (1)}_{1j(k)} - X^1_{|r} C^{r(1)}_{j(k)} - X^1|^{(1)}_{(r)} P^{(r)\ (1)}_{(1)j(k)}, \\[6pt] X^1|^{(1)}_{(j)}|^{(1)}_{(k)} - X^1|^{(1)}_{(k)}|^{(1)}_{(j)} = X^1 \bar{S}^{1(1)(1)}_{1(j)(k)} - X^1|^{(1)}_{(r)} S^{(r)(1)(1)}_{(1)(j)(k)}, \end{cases}$$

$$(h_M) \begin{cases} X^i_{/1|k} - X^i_{|k/1} = X^r R^i_{r1k} - X^i_{/1} \bar{T}^1_{1k} - X^i_{|r} T^r_{1k} - X^i|^{(1)}_{(r)} R^{(r)}_{(1)1k}, \\[6pt] X^i_{|j|k} - X^i_{|k|j} = X^r R^i_{rjk} - X^i_{|r} T^r_{jk} - X^i|^{(1)}_{(r)} R^{(r)}_{(1)jk}, \\[6pt] X^i_{/1}|^{(1)}_{(k)} - X^i|^{(1)}_{(k)/1} = X^r P^{i\ (1)}_{r1(k)} - X^i_{/1} \bar{C}^{1(1)}_{1(k)} - X^i|^{(1)}_{(r)} P^{(r)\ (1)}_{(1)1(k)}, \\[6pt] X^i_{|j}|^{(1)}_{(k)} - X^i|^{(1)}_{(k)|j} = X^r P^{i\ (1)}_{rj(k)} - X^i_{|r} C^{r(1)}_{j(k)} - X^i|^{(1)}_{(r)} P^{(r)\ (1)}_{(1)j(k)}, \\[6pt] X^i|^{(1)}_{(j)}|^{(1)}_{(k)} - X^i|^{(1)}_{(k)}|^{(1)}_{(j)} = X^r S^{i(1)(1)}_{r(j)(k)} - X^i|^{(1)}_{(r)} S^{(r)(1)(1)}_{(1)(j)(k)}, \end{cases}$$

$$(v) \begin{cases} X^{(i)}_{(1)/1|k} - X^{(i)}_{(1)|k/1} = X^{(r)}_{(1)} R^{(i)(1)}_{(1)(r)1k} - X^{(i)}_{(1)/1} \bar{T}^1_{1k} - X^{(i)}_{(1)|r} T^r_{1k} \\[4pt] \qquad\qquad\qquad - X^{(i)}_{(1)}|^{(1)}_{(r)} R^{(r)}_{(1)1k}, \\[8pt] X^{(i)}_{(1)|j|k} - X^{(i)}_{(1)|k|j} = X^{(r)}_{(1)} R^{(i)(1)}_{(1)(r)jk} - X^{(i)}_{(1)|r} T^r_{jk} - X^{(i)}_{(1)}|^{(1)}_{(r)} R^{(r)}_{(1)jk}, \\[8pt] X^{(i)}_{(1)/1}|^{(1)}_{(k)} - X^{(i)}_{(1)}|^{(1)}_{(k)/1} = X^{(r)}_{(1)} P^{(i)(1)\ (1)}_{(1)(r)1(k)} - X^{(i)}_{(1)/1} \bar{C}^{1(1)}_{1(k)} \\[4pt] \qquad\qquad\qquad - X^{(i)}_{(1)}|^{(1)}_{(r)} P^{(r)\ (1)}_{(1)1(k)}, \\[8pt] X^{(i)}_{(1)|j}|^{(1)}_{(k)} - X^{(i)}_{(1)}|^{(1)}_{(k)|j} = X^{(r)}_{(1)} P^{(i)(1)\ (1)}_{(1)(r)j(k)} - X^{(i)}_{(1)|r} C^{r(1)}_{j(k)} \\[4pt] \qquad\qquad\qquad - X^{(i)}_{(1)}|^{(1)}_{(r)} P^{(r)\ (1)}_{(1)j(k)}, \\[8pt] X^{(i)}_{(1)}|^{(1)}_{(j)}|^{(1)}_{(k)} - X^{(i)}_{(1)}|^{(1)}_{(k)}|^{(1)}_{(j)} = X^{(r)}_{(1)} S^{(i)(1)(1)(1)}_{(1)(r)(j)(k)} - X^{(i)}_{(1)}|^{(1)}_{(r)} S^{(r)(1)(1)}_{(1)(j)(k)}. \end{cases}$$

Proof: Let (Y_A) and (ω^A), where $A \in \left\{ 1, i_{,(i)}^{(1)} \right\}$, be the dual bases adapted to the nonlinear connection Γ and let $X = X^F Y_F$ be a distinguished vector field on the 1-jet space $E = J^1(\mathbb{R}, M)$. In this context, using the equalities

1. $\nabla_{Y_C} Y_B = \Gamma_{BC}^F Y_F,$

2. $[Y_B, Y_C] = R_{BC}^F Y_F,$

3. $\mathbf{T}(Y_C, Y_B) = \mathbf{T}_{BC}^F Y_F = \{\Gamma_{BC}^F - \Gamma_{CB}^F - R_{CB}^F\} Y_F,$

4. $\mathbf{R}(Y_C, Y_B) Y_A = \mathbf{R}_{ABC}^F Y_F,$

5. $\nabla_{Y_C} \omega^B = -\Gamma_{FC}^B \omega^F,$

6. $[\mathbf{R}(Y_C, Y_B) X] \cdot \omega^B \cdot \omega^C = \{\nabla_{Y_C} \nabla_{Y_B} X - \nabla_{Y_B} \nabla_{Y_C} X$
    $$-\nabla_{[Y_C, Y_B]} X\} \cdot \omega^B \cdot \omega^C,$$

where " $\cdot$ " represents the tensorial product "$\otimes$," we deduce by a direct calculation that

$$X_{:B:C}^A - X_{:C:B}^A = X^F \mathbf{R}_{FBC}^A - X_{:F}^A \mathbf{T}_{BC}^F, \tag{2.4}$$

where " $_{:D}$ " represents one from the local covariant derivatives "$_{/1}$," "$_{|j}$," or "$|_{(j)}^{(1)}$" produced by the Γ-linear connection ∇.

Taking into account that the indices $A, B, C, \ldots$ belong to the set $\left\{ 1, i_{,(i)}^{(1)} \right\}$, by complicated local computations, the identities (2.4) imply the required Ricci identities.

Now, let us consider the canonical Liouville d-tensor field

$$\mathbf{C} = \mathbf{C}_{(1)}^{(i)} \frac{\partial}{\partial y_1^i} = y_1^i \frac{\partial}{\partial y_1^i}.$$

Definition 48. The distinguished tensors defined by the local components

$$\bar{D}_{(1)1}^{(i)} = \mathbf{C}_{(1)/1}^{(i)}, \quad D_{(1)j}^{(i)} = \mathbf{C}_{(1)|j}^{(i)}, \quad d_{(1)(j)}^{(i)(1)} = \mathbf{C}_{(1)}^{(i)}|_{(j)}^{(1)} \tag{2.5}$$

are called the *nonmetrical deflection d-tensors attached to the Γ-linear connection ∇* on the 1-jet space $E = J^1(\mathbb{R}, M)$.

Taking into account the expressions of the local covariant derivatives of the Γ-linear connection ∇ given by (2.2), by a direct calculation, we find the following:

Proposition 49. *The nonmetrical deflection d-tensors of the Γ-linear connection ∇ have the expressions*

$$\bar{D}_{(1)1}^{(i)} = -M_{(1)1}^{(i)} + G_{(1)(r)1}^{(i)(1)} y_1^r, \quad D_{(1)j}^{(i)} = -N_{(1)j}^{(i)} + L_{(1)(r)j}^{(i)(1)} y_1^r,$$

$$\tag{2.6}$$

$$d_{(1)(j)}^{(i)(1)} = \delta_j^i + C_{(1)(r)(j)}^{(i)(1)(1)} y_1^r.$$

In the sequel, applying the set (v) of the Ricci identities to the components of the canonical Liouville d-tensor field $\mathbf{C}$, we get

Theorem 50. *The nonmetrical deflection d-tensors attached to the Γ-linear connection ∇ on the 1-jet space $E = J^1(\mathbb{R}, M)$ verify the following identities:*

$$
\begin{cases}
\bar{D}^{(i)}_{(1)1|k} - D^{(i)}_{(1)k/1} = y_1^r R^{(i)(1)}_{(1)(r)1k} - \bar{D}^{(i)}_{(1)1} \bar{T}^1_{1k} - D^{(i)}_{(1)r} T^r_{1k} - d^{(i)(1)}_{(1)(r)} R^{(r)}_{(1)1k}, \\[2mm]
D^{(i)}_{(1)j|k} - D^{(i)}_{(1)k|j} = y_1^r R^{(i)(1)}_{(1)(r)jk} - D^{(i)}_{(1)r} T^r_{jk} - d^{(i)(1)}_{(1)(r)} R^{(r)}_{(1)jk}, \\[2mm]
\bar{D}^{(i)}_{(1)1}|^{(1)}_{(k)} - d^{(i)(1)}_{(1)(k)/1} = y_1^r P^{(i)(1)}_{(1)(r)1(k)}{}^{(1)} - \bar{D}^{(i)}_{(1)1} \bar{C}^{1(1)}_{1(k)} - d^{(i)(1)}_{(1)(r)} P^{(r)}_{(1)1(k)}{}^{(1)}, \\[2mm]
D^{(i)}_{(1)j}|^{(1)}_{(k)} - d^{(i)(1)}_{(1)(k)|j} = y_1^r P^{(i)(1)}_{(1)(r)j(k)}{}^{(1)} - D^{(i)}_{(1)r} C^{r(1)}_{j(k)} - d^{(i)(1)}_{(1)(r)} P^{(r)}_{(1)j(k)}{}^{(1)}, \\[2mm]
d^{(i)(1)}_{(1)(j)}|^{(1)}_{(k)} - d^{(i)(1)}_{(1)(k)}|^{(1)}_{(j)} = y_1^r S^{(i)(1)(1)(1)}_{(1)(r)(j)(k)} - d^{(i)(1)}_{(1)(r)} S^{(r)(1)(1)}_{(1)(j)(k)}.
\end{cases}
\tag{2.7}
$$

CHAPTER 3

LOCAL BIANCHI IDENTITIES IN THE RELATIVISTIC TIME-DEPENDENT LAGRANGE GEOMETRY

Because of the huge number of nontrivial adapted components (nine sets) which characterize a general Γ-linear connection ∇, the aim of this Chapter is to introduce the geometrical concept of *h-normal Γ-linear connection* on the 1-jet space $J^1(\mathbb{R}, M)$, whose local adapted components reduce only to *four sets*. This new geometrical object naturally generalizes the notion of *normal N-linear connection* from the theory of Finsler spaces [20] or Lagrange spaces [55] on the tangent bundle.

On the other hand, the importance of Bianchi identities in the geometrization of time-dependent Lagrangians is well known (see Miron [55], [56], Crampin [29], or Neagu [63]). The present Chapter describes the local Bianchi identities attached to an h-normal Γ-linear connection of Cartan type.

3.1 THE ADAPTED COMPONENTS OF h-NORMAL Γ-LINEAR CONNECTIONS

Let us suppose that the 1-jet space $E = J^1(\mathbb{R}, M)$ is endowed with a nonlinear connection

$$\Gamma = \left(M^{(i)}_{(1)1}, N^{(i)}_{(1)j} \right). \tag{3.1}$$

Let us consider also that we have on the 1-jet space E a fixed Γ-linear connection ∇, defined by the adapted local coefficients

$$\nabla\Gamma = \left(\bar{G}_{11}^1,\ G_{i1}^k,\ G_{(1)(j)1}^{(i)(1)},\ \bar{L}_{1j}^1,\ L_{ij}^k,\ L_{(1)(j)k}^{(i)(1)},\ \bar{C}_{1(k)}^{1(1)},\ C_{i(k)}^{j(1)},\ C_{(1)(j)(k)}^{(i)(1)(1)}\right). \tag{3.2}$$

Definition 51. A Γ-linear connection ∇ on E, whose local components (3.2) verify the relations

$$\bar{G}_{11}^1 = \kappa_{11}^1, \quad \bar{L}_{1j}^1 = 0, \quad \bar{C}_{1(k)}^{1(1)} = 0, \quad \nabla\mathbf{J}_h = 0,$$

where $h = (h_{11}(t))$ is a Riemannian metric on $\mathbb{R}$, κ_{11}^1 is its Christoffel symbol, and $\mathbf{J}_h$ is the h-normalization d-tensor field (see Example 1.3), is called an h-*normal* Γ-*linear connection* on $E = J^1(\mathbb{R}, M)$.

Remark 52. The condition $\nabla\mathbf{J}_h = 0$ is equivalent with the local equalities

$$J_{(1)1j/1}^{(i)} = 0, \quad J_{(1)1j|k}^{(i)} = 0, \quad J_{(1)1j}^{(i)}\big|_{(k)}^{(1)} = 0,$$

where "$/_1$," "$|_k$," and "$\big|_{(k)}^{(1)}$" represent the $\mathbb{R}$-horizontal, M-horizontal, and vertical local covariant derivatives produced by the Γ-linear connection $\nabla\Gamma$.

In this context, we can prove the following important local geometrical result:

Theorem 53. *The components of an h-normal Γ-linear connection ∇ verify the identities*

$$\begin{aligned}
\bar{G}_{11}^1 &= \kappa_{11}^1, & \bar{L}_{1j}^1 &= 0, & \bar{C}_{1(k)}^{1(1)} &= 0, \\
G_{(1)(i)1}^{(k)(1)} &= G_{i1}^k - \delta_i^k\kappa_{11}^1, & L_{(1)(i)j}^{(k)(1)} &= L_{ij}^k, & C_{(1)(i)(j)}^{(k)(1)(1)} &= C_{i(j)}^{k(1)}.
\end{aligned} \tag{3.3}$$

Proof: The first three relations from (3.3) are a direct consequence of the definition of an h-normal Γ-linear connection ∇.

The condition $\nabla\mathbf{J}_h = 0$ implies the local relations

$$h_{11}G_{(1)(j)1}^{(i)(1)} = h_{11}G_{j1}^i + \delta_j^i\left[\kappa_{111} - \frac{dh_{11}}{dt}\right],$$

$$h_{11}L_{(1)(j)k}^{(i)(1)} = h_{11}L_{jk}^i, \quad h_{11}C_{(1)(j)(k)}^{(i)(1)(1)} = h_{11}C_{j(k)}^{i(1)},$$

where $\kappa_{111} = \kappa_{11}^1 h_{11}$ represent the Christoffel symbols of first kind attached to the Riemannian metric $h_{11}(t)$. Contracting the above relations with the inverse $h^{11} = 1/h_{11}$, we obtain the last three identities from (3.3). $\blacksquare$

Remark 54. Theorem 53 implies that an h-normal Γ-linear connection ∇ is determined by *four* effective local components (instead of *two* effective local components for a normal N-linear connection in the Miron's case [56], p. 21), namely,

$$\nabla\Gamma = \left(\kappa_{11}^1, G_{i1}^k, L_{ij}^k, C_{i(j)}^{k(1)}\right). \tag{3.4}$$

The other five components of $\nabla\Gamma$ cancel or depend on the above four components, via the formulas (3.3).

■ **EXAMPLE 3.1**

The canonical Berwald $\overset{\circ}{\Gamma}$-linear connection associated to the pair of metrics $(h_{11}(t), \varphi_{ij}(x))$ is an h-normal $\overset{\circ}{\Gamma}$-linear connection, defined by the local components

$$B\overset{\circ}{\Gamma} = \left(\kappa_{11}^1, 0, \gamma_{ij}^k, 0 \right).$$

The study of adapted components of the torsion tensor $\mathbf{T}$ and curvature tensor $\mathbf{R}$ of an arbitrary Γ-linear connection ∇ on $E = J^1(\mathbb{R}, M)$ was completely done in the preceding Chapter, where we proved that the torsion tensor $\mathbf{T}$ is determined by *ten* effective local d-tensors, while the curvature tensor $\mathbf{R}$ is determined by *fifteen* effective local d-tensors. In the sequel, we study the adapted components of the torsion and curvature tensors for an h-normal Γ-linear connection $\nabla\Gamma$ given by (3.4) and (3.3).

Theorem 55. *The torsion tensor $\mathbf{T}$ of an h-normal Γ-linear connection ∇ on E is determined only by the following* **eight** *adapted local d-tensors (instead of* **ten** *in the general case):*

		$h_{\mathbb{R}}$	h_M	v
$h_{\mathbb{R}} h_{\mathbb{R}}$		0	0	0
$h_M h_{\mathbb{R}}$		0	T_{1j}^r	$R_{(1)1j}^{(r)}$
$h_M h_M$		0	T_{ij}^r	$R_{(1)ij}^{(r)}$
$v h_{\mathbb{R}}$		0	0	$P_{(1)1(j)}^{(r)\,(1)}$
$v h_M$		0	$P_{i(j)}^{r(1)}$	$P_{(1)i(j)}^{(r)\,(1)}$
$v v$		0	0	$S_{(1)(i)(j)}^{(r)(1)(1)}$

(3.5)

where

1. $T_{1j}^r = -G_{j1}^r$,

2. $R_{(1)1j}^{(r)} = \dfrac{\delta M_{(1)1}^{(r)}}{\delta x^j} - \dfrac{\delta N_{(1)j}^{(r)}}{\delta t}$,

3. $T_{ij}^r = L_{ij}^r - L_{ji}^r$,

4. $R_{(1)ij}^{(r)} = \dfrac{\delta N_{(1)i}^{(r)}}{\delta x^j} - \dfrac{\delta N_{(1)j}^{(r)}}{\delta x^i}$,

5. $P_{(1)1(j)}^{(r)\,(1)} = \dfrac{\partial M_{(1)1}^{(r)}}{\partial y_1^j} - G_{j1}^r + \delta_j^r \kappa_{11}^1$,

6. $P_{i(j)}^{r(1)} = C_{i(j)}^{r(1)}$,

7. $P_{(1)i(j)}^{(r)\,(1)} = \dfrac{\partial N_{(1)i}^{(r)}}{\partial y_1^j} - L_{ji}^r,$

8. $S_{(1)(i)(j)}^{(r)(1)(1)} = C_{i(j)}^{r(1)} - C_{j(i)}^{r(1)}.$

Proof: Particularizing the general local expressions from Theorem 43 (which give those *ten* components of the torsion tensor of a Γ-linear connection ∇, in the large) for an h-normal Γ-linear connection ∇, we deduce that the adapted local components $\bar{T}_{1j}^1$ and $\bar{P}_{1(j)}^{1(1)}$ vanish, while the other *eight* ones from Table 3.5 are expressed by the preceding formulas.

Remark 56. The torsion of a normal N-linear connection in the Miron's case is characterized only by *five* effective adapted components (see [56], p. 24).

Remark 57. For the Berwald $\mathring{\Gamma}$-linear connection $B\mathring{\Gamma}$, associated to the metrics $h_{11}(t)$ and $\varphi_{ij}(x)$, all adapted local torsion d-tensors vanish, except

$$R_{(1)ij}^{(k)} = \mathfrak{R}_{mij}^k y_1^m,$$

where $\mathfrak{R}_{mij}^k(x)$ are the classical local curvature components of the semi-Riemannian metric $\varphi_{ij}(x)$.

The expressions of the local curvature d-tensors of an arbitrary Γ-linear connection, together with the particular properties of an h-normal Γ-linear connection, imply a considerable reduction (from *fifteen* to *five*) of the <u>effective</u> local curvature d-tensors that characterize an h-normal Γ-linear connection. In other words, we have the following:

Theorem 58. *The curvature tensor* **R** *of an h-normal Γ-linear connection ∇ on E is characterized by* **five** *effective local curvature d-tensors (instead of* **fifteen** *in the general case):*

		$h_{\mathbb{R}}$	h_M	v
$h_{\mathbb{R}}h_{\mathbb{R}}$		0	0	0
$h_M h_{\mathbb{R}}$		0	R_{i1k}^l	$R_{(1)(i)1k}^{(l)(1)} = R_{i1k}^l$
$h_M h_M$		0	R_{ijk}^l	$R_{(1)(i)jk}^{(l)(1)} = R_{ijk}^l$
$vh_{\mathbb{R}}$		0	$P_{i1(k)}^{l\ (1)}$	$P_{(1)(i)1(k)}^{(l)(1)\ (1)} = P_{i1(k)}^{l\ (1)}$
vh_M		0	$P_{ij(k)}^{l\ (1)}$	$P_{(1)(i)j(k)}^{(l)(1)\ (1)} = P_{ij(k)}^{l\ (1)}$
vv		0	$S_{i(j)(k)}^{l(1)(1)}$	$S_{(1)(i)(j)(k)}^{(l)(1)(1)(1)} = S_{i(j)(k)}^{l(1)(1)}$

(3.6)

where

1. $R_{i1k}^l = \dfrac{\delta G_{i1}^l}{\delta x^k} - \dfrac{\delta L_{ik}^l}{\delta t} + G_{i1}^r L_{rk}^l - L_{ik}^r G_{r1}^l + C_{i(r)}^{l(1)} R_{(1)1k}^{(r)},$

2. $R^l_{ijk} = \dfrac{\delta L^l_{ij}}{\delta x^k} - \dfrac{\delta L^l_{ik}}{\delta x^j} + L^r_{ij} L^l_{rk} - L^r_{ik} L^l_{rj} + C^{l(1)}_{i(r)} R^{(r)}_{(1)jk},$

3. $P^{l\ (1)}_{i1(k)} = \dfrac{\partial G^l_{i1}}{\partial y^k_1} - C^{l(1)}_{i(k)/1} + C^{l(1)}_{i(r)} P^{(r)\ (1)}_{(1)1(k)},$

4. $P^{l\ (1)}_{ij(k)} = \dfrac{\partial L^l_{ij}}{\partial y^k_1} - C^{l(1)}_{i(k)|j} + C^{l(1)}_{i(r)} P^{(r)\ (1)}_{(1)j(k)},$

5. $S^{l(1)(1)}_{i(j)(k)} = \dfrac{\partial C^{l(1)}_{i(j)}}{\partial y^k_1} - \dfrac{\partial C^{l(1)}_{i(k)}}{\partial y^j_1} + C^{r(1)}_{i(j)} C^{l(1)}_{r(k)} - C^{r(1)}_{i(k)} C^{l(1)}_{r(j)}.$

Proof: The general formulas that express those *fifteen* local curvature d-tensors of an arbitrary Γ-linear connection (see Theorem 45 and Theorem 46), applied to the particular case of an h-normal Γ-linear connection ∇ on $E = J^1(\mathbb{R}, M)$, imply the preceding *five* formulas and the relations from Table 3.6.

Remark 59. The curvature of a normal N-linear connection in the Miron's case is characterized only by *three* effective adapted components (see [56], p. 25).

Remark 60. For the Berwald $\overset{\circ}{\Gamma}$-linear connection $B\overset{\circ}{\Gamma}$, associated to the metrics $h_{11}(t)$ and $\varphi_{ij}(x)$, all local curvature d-tensors vanish, except

$$R^{(l)(1)}_{(1)(i)jk} = R^l_{ijk} = \mathfrak{R}^l_{ijk},$$

where $\mathfrak{R}^l_{ijk}(x)$ are the local curvature tensors of the semi-Riemannian metric $\varphi_{ij}(x)$.

3.2 DEFLECTION d-TENSOR IDENTITIES AND LOCAL BIANCHI IDENTITIES FOR d-CONNECTIONS OF CARTAN TYPE

Because of the reduced number and the simplified form of the local torsion and curvature d-tensors of an h-normal Γ-linear connection ∇ on the 1-jet space E, the number of attached *local Ricci and Bianchi identities* considerably simplifies. A substantial reduction of these identities obtains considering the more particular case of an h-*normal* Γ-*linear connection of Cartan type*.

Definition 61. An h-normal Γ-linear connection on $E = J^1(\mathbb{R}, M)$, whose local components

$$\nabla\Gamma = \left(\kappa^1_{11}, G^k_{i1}, L^k_{ij}, C^{k(1)}_{i(j)} \right)$$

verify the supplementary conditions $L^k_{ij} = L^k_{ji}$ and $C^{k(1)}_{i(j)} = C^{k(1)}_{j(i)}$, is called an h-*normal* Γ-*linear connection of Cartan type*.

Remark 62. In the particular case of an h-normal Γ-linear connection of Cartan type, the conditions $L^k_{ij} = L^k_{ji}$ and $C^{k(1)}_{i(j)} = C^{k(1)}_{j(i)}$ imply the torsion equalities

$$T^k_{ij} = 0, \quad S^{(k)(1)(1)}_{(1)(i)(j)} = 0.$$

Rewriting the local Ricci identities of a Γ-linear connection ∇ (described in detail in Theorem 47), for the particular case of an h-normal Γ-linear connection of Cartan type, we find a simplified form of these identities. Consequently, we obtain the following:

Theorem 63. *The following* **local Ricci identities** *for an h-normal Γ-linear connection of Cartan type are true:*

$$
(h_{\mathbb{R}}) \begin{cases}
X^1_{/1|k} - X^1_{|k/1} = -X^1_{|r}T^r_{1k} - X^1|^{(1)}_{(r)}R^{(r)}_{(1)1k}, \\[4pt]
X^1_{|j|k} - X^1_{|k|j} = -X^1|^{(1)}_{(r)}R^{(r)}_{(1)jk}, \\[4pt]
X^1_{/1}|^{(1)}_{(k)} - X^1|^{(1)}_{(k)/1} = -X^1|^{(1)}_{(r)}P^{(r)\ (1)}_{(1)1(k)}, \\[4pt]
X^1_{|j}|^{(1)}_{(k)} - X^1|^{(1)}_{(k)|j} = -X^1_{|r}C^{r(1)}_{j(k)} - X^1|^{(1)}_{(r)}P^{(r)\ (1)}_{(1)j(k)}, \\[4pt]
X^1|^{(1)}_{(j)}|^{(1)}_{(k)} - X^1|^{(1)}_{(k)}|^{(1)}_{(j)} = 0,
\end{cases}
$$

$$
(h_M) \begin{cases}
X^i_{/1|k} - X^i_{|k/1} = X^r R^i_{r1k} - X^i_{|r}T^r_{1k} - X^i|^{(1)}_{(r)}R^{(r)}_{(1)1k}, \\[4pt]
X^i_{|j|k} - X^i_{|k|j} = X^r R^i_{rjk} - X^i|^{(1)}_{(r)}R^{(r)}_{(1)jk}, \\[4pt]
X^i_{/1}|^{(1)}_{(k)} - X^i|^{(1)}_{(k)/1} = X^r P^{i\ (1)}_{r1(k)} - X^i|^{(1)}_{(r)}P^{(r)\ (1)}_{(1)1(k)}, \\[4pt]
X^i_{|j}|^{(1)}_{(k)} - X^i|^{(1)}_{(k)|j} = X^r P^{i\ (1)}_{rj(k)} - X^i_{|r}C^{r(1)}_{j(k)} - X^i|^{(1)}_{(r)}P^{(r)\ (1)}_{(1)j(k)}, \\[4pt]
X^i|^{(1)}_{(j)}|^{(1)}_{(k)} - X^i|^{(1)}_{(k)}|^{(1)}_{(j)} = X^r S^{i(1)(1)}_{r(j)(k)},
\end{cases}
$$

$$
(v) \begin{cases}
X^{(i)}_{(1)/1|k} - X^{(i)}_{(1)|k/1} = X^{(r)}_{(1)}R^i_{r1k} - X^{(i)}_{(1)|r}T^r_{1k} - X^{(i)}_{(1)}|^{(1)}_{(r)}R^{(r)}_{(1)1k}, \\[4pt]
X^{(i)}_{(1)|j|k} - X^{(i)}_{(1)|k|j} = X^{(r)}_{(1)}R^i_{rjk} - X^{(i)}_{(1)}|^{(1)}_{(r)}R^{(r)}_{(1)jk}, \\[4pt]
X^{(i)}_{(1)/1}|^{(1)}_{(k)} - X^{(i)}_{(1)}|^{(1)}_{(k)/1} = X^{(r)}_{(1)}P^{i\ (1)}_{r1(k)} - X^{(i)}_{(1)}|^{(1)}_{(r)}P^{(r)\ (1)}_{(1)1(k)}, \\[4pt]
X^{(i)}_{(1)|j}|^{(1)}_{(k)} - X^{(i)}_{(1)}|^{(1)}_{(k)|j} = X^{(r)}_{(1)}P^{i\ (1)}_{rj(k)} - X^{(i)}_{(1)|r}C^{r(1)}_{j(k)} - X^{(i)}_{(1)}|^{(1)}_{(r)}P^{(r)\ (1)}_{(1)j(k)}, \\[4pt]
X^{(i)}_{(1)}|^{(1)}_{(j)}|^{(1)}_{(k)} - X^{(i)}_{(1)}|^{(1)}_{(k)}|^{(1)}_{(j)} = X^{(r)}_{(1)}S^{i(1)(1)}_{r(j)(k)},
\end{cases}
$$

where

$$
X = X^1\frac{\delta}{\delta t} + X^i\frac{\delta}{\delta x^i} + X^{(i)}_{(1)}\frac{\partial}{\partial y^i_1}
$$

is an arbitrary distinguished vector field on the 1-jet space $E = J^1(\mathbb{R}, M)$.

In what follows, let us construct the *nonmetrical deflection d-tensors*

$$
\bar{D}^{(i)}_{(1)1} = C^{(i)}_{(1)/1} = y^i_{1/1}, \quad D^{(i)}_{(1)j} = C^{(i)}_{(1)|j} = y^i_{1|j}, \quad d^{(i)(1)}_{(1)(j)} = C^{(i)}_{(1)}|^{(1)}_{(j)} = y^i_1|^{(1)}_{(j)},
$$

attached to the h-normal Γ-linear connection (3.4). Then, a direct calculation yields the following:

Proposition 64. *The nonmetrical deflection d-tensors attached to the h-normal Γ-linear connection ∇ given by (3.4) have the expressions*

$$\bar{D}^{(i)}_{(1)1} = -M^{(i)}_{(1)1} + G^i_{r1}y^r_1 - \kappa^1_{11}y^i_1, \qquad D^{(i)}_{(1)j} = -N^{(i)}_{(1)j} + L^i_{rj}y^r_1,$$

$$d^{(i)(1)}_{(1)(j)} = \delta^i_j + C^{i(1)}_{r(j)}y^r_1. \tag{3.7}$$

Applying now the general nonmetrical deflection d-tensor identities (2.7) to the h-normal Γ-linear connection of Cartan type ∇ given by (3.4), we find the following:

Theorem 65. *The following* **five** *identities of the nonmetrical deflection d-tensors associated to an h-normal Γ-linear connection of Cartan type (instead of* **three** *in the Miron's case [56], p. 80) are true:*

$$\begin{cases} \bar{D}^{(i)}_{(1)1|k} - D^{(i)}_{(1)k/1} = y^r_1 R^i_{r1k} - D^{(i)}_{(1)r}T^r_{1k} - d^{(i)(1)}_{(1)(r)}R^{(r)}_{(1)1k}, \\[2mm] D^{(i)}_{(1)j|k} - D^{(i)}_{(1)k|j} = y^r_1 R^i_{rjk} - d^{(i)(1)}_{(1)(r)}R^{(r)}_{(1)jk}, \\[2mm] \bar{D}^{(i)}_{(1)1}|^{(1)}_{(k)} - d^{(i)(1)}_{(1)(k)/1} = y^r_1 P^{i\ (1)}_{r1(k)} - d^{(i)(1)}_{(1)(r)}P^{(r)\ (1)}_{(1)1(k)}, \\[2mm] D^{(i)}_{(1)j}|^{(1)}_{(k)} - d^{(i)(1)}_{(1)(k)|j} = y^r_1 P^{i\ (1)}_{rj(k)} - D^{(i)}_{(1)r}C^{r(1)}_{j(k)} - d^{(i)(1)}_{(1)(r)}P^{(r)\ (1)}_{(1)j(k)}, \\[2mm] d^{(i)(1)}_{(1)(j)}|^{(1)}_{(k)} - d^{(i)(1)}_{(1)(k)}|^{(1)}_{(j)} = y^r_1 S^{i(1)(1)}_{r(j)(k)}. \end{cases} \tag{3.8}$$

Remark 66. The identities (3.8) are used in the local description of *geometrical Maxwell equations* that govern the *jet single-time electromagnetic 2-form* produced by a relativistic time-dependent Lagrangian on the 1-jet space $E = J^1(\mathbb{R}, M)$.

The use of h-normal Γ-linear connections of Cartan type in the study of differential geometry of the 1-jet vector bundle $E = J^1(\mathbb{R}, M)$ is also convenient because the number and the form of the local Bianchi identities associated to such connections are considerably simplified. In fact, we have the following:

Theorem 67. *The following* **nineteen** *effective* **local Bianchi identities** *for the h-normal Γ-linear connection of Cartan type ∇ given by (3.4) are true:*

1. $\mathcal{A}_{\{j,k\}} \left\{ R^l_{j1k} + T^l_{1j|k} + R^{(r)}_{(1)1j}C^{l(1)}_{k(r)} \right\} = 0,$

2*. $\sum_{\{i,j,k\}} \left\{ R^l_{ijk} - R^{(r)}_{(1)ij}C^{l(1)}_{k(r)} \right\} = 0,$

3. $\mathcal{A}_{\{j,k\}} \left\{ R^{(l)}_{(1)1j|k} + T^r_{1j}R^{(l)}_{(1)kr} + R^{(r)}_{(1)1j}P^{(l)\ (1)}_{(1)k(r)} \right\}$

$$= -R^{(l)}_{(1)jk/1} - R^{(r)}_{(1)jk}P^{(l)\ (1)}_{(1)1(r)},$$

4*. $\sum_{\{i,j,k\}} \left\{ R^{(l)}_{(1)ij|k} + R^{(r)}_{(1)ij}P^{(l)\ (1)}_{(1)k(r)} \right\} = 0,$

5. $T^l_{1k}\big|^{(1)}_{(p)} - C^{l(1)}_{r(p)} T^r_{1k} + P^l_{k1(p)}{}^{(1)} + C^{l(1)}_{k(p)/1} + C^{r(1)}_{k(p)} T^l_{1r} - C^{l(1)}_{k(r)} P^{(r)}_{(1)1(p)}{}^{(1)} = 0,$

6[*]. $\mathcal{A}_{\{j,k\}} \left\{ C^{l(1)}_{j(p)|k} + C^{l(1)}_{k(r)} P^{(r)}_{(1)j(p)}{}^{(1)} + P^l_{jk(p)}{}^{(1)} \right\} = 0,$

7. $P^{(l)}_{(1)1(p)|k}{}^{(1)} - P^{(l)}_{(1)k(p)/1}{}^{(1)} + P^{(l)}_{(1)k(r)}{}^{(1)} P^{(r)}_{(1)1(p)}{}^{(1)} - P^{(l)}_{(1)1(r)}{}^{(1)} P^{(r)}_{(1)k(p)}{}^{(1)}$

$$= R^{(l)}_{(1)1k}\big|^{(1)}_{(p)} - R^l_{p1k} + R^{(l)}_{(1)1r} C^{r(1)}_{k(p)} - T^r_{1k} P^{(l)}_{(1)r(p)}{}^{(1)},$$

8[*]. $\mathcal{A}_{\{j,k\}} \left\{ R^{(l)}_{(1)jr} C^{r(1)}_{k(p)} + P^{(l)}_{(1)j(r)}{}^{(1)} P^{(r)}_{(1)k(p)}{}^{(1)} + P^{(l)}_{(1)k(p)|j}{}^{(1)} \right\} = R^l_{pjk} - R^{(l)}_{(1)jk}\big|^{(1)}_{(p)},$

9[*]. $\mathcal{A}_{\{j,k\}} \left\{ C^{l(1)}_{i(j)}\big|^{(1)}_{(k)} + C^{r(1)}_{i(k)} C^{l(1)}_{r(j)} \right\}$

$$= \mathcal{A}_{\{j,k\}} \left\{ \frac{C^{l(1)}_{i(j)}}{\partial y^k_1} + C^{r(1)}_{i(j)} C^{l(1)}_{r(k)} \right\} = S^{l(1)(1)}_{i(j)(k)},$$

10. $\mathcal{A}_{\{j,k\}} \left\{ P^{(l)}_{(1)1(j)}{}^{(1)}\big|^{(1)}_{(k)} + P^l_{j1(k)}{}^{(1)} \right\} = 0,$

11[*]. $\mathcal{A}_{\{j,k\}} \left\{ P^l_{ji(k)}{}^{(1)} + P^{(l)}_{(1)r(j)}{}^{(1)} C^{r(1)}_{i(k)} - P^{(l)}_{(1)i(k)}{}^{(1)}\big|^{(1)}_{(j)} \right\} = 0,$

12[*]. $\sum_{\{i,j,k\}} S^{l(1)(1)}_{i(j)(k)} = 0,$

13. $\mathcal{A}_{\{j,k\}} \left\{ R^l_{p1j|k} + T^r_{1j} R^l_{pkr} + R^{(r)}_{(1)1j} P^l_{pk(r)}{}^{(1)} \right\} = -R^l_{pjk/1} - R^{(r)}_{(1)jk} P^l_{p1(r)}{}^{(1)},$

14[*]. $\sum_{\{i,j,k\}} \left\{ R^l_{pij|k} + R^{(r)}_{(1)ij} P^l_{pk(r)}{}^{(1)} \right\} = 0,$

15. $P^l_{i1(p)|k}{}^{(1)} - P^l_{ik(p)/1}{}^{(1)} + P^{(r)}_{(1)1(p)}{}^{(1)} P^l_{ik(r)}{}^{(1)} - P^{(r)}_{(1)k(p)}{}^{(1)} P^l_{i1(r)}{}^{(1)}$

$$= R^l_{i1k}\big|^{(1)}_{(p)} + R^{(r)}_{(1)1k} S^{l(1)(1)}_{i(p)(r)} + C^{r(1)}_{k(p)} R^l_{i1r} - T^r_{1k} P^l_{ir(p)}{}^{(1)},$$

16[*]. $\mathcal{A}_{\{j,k\}} \left\{ R^l_{ijr} C^{r(1)}_{k(p)} + P^l_{ij(r)}{}^{(1)} P^{(r)}_{(1)k(p)}{}^{(1)} + P^l_{ik(p)|j}{}^{(1)} \right\}$

$$= -S^{l(1)(1)}_{i(p)(r)} R^{(r)}_{(1)jk} - R^l_{ijk}\big|^{(1)}_{(p)},$$

17[*]. $\mathcal{A}_{\{j,k\}} \left\{ P^l_{p1(j)}{}^{(1)}\big|^{(1)}_{(k)} + P^{(r)}_{(1)1(j)}{}^{(1)} S^{l(1)(1)}_{p(k)(r)} \right\} = -S^{l(1)(1)}_{p(j)(k)/1},$

18[*]. $\mathcal{A}_{\{j,k\}} \left\{ P^l_{pr(j)}{}^{(1)} C^{r(1)}_{i(k)} - S^{l(1)(1)}_{p(j)(r)} P^{(r)}_{(1)i(k)}{}^{(1)} - P^l_{pi(k)}{}^{(1)}\big|^{(1)}_{(j)} \right\} = -S^{l(1)(1)}_{p(j)(k)|i},$

19[*]. $\sum_{\{i,j,k\}} S^{l(1)(1)}_{p(i)(j)}\big|^{(1)}_{(k)} = 0,$

where $\sum_{\{i,j,k\}}$ *represents a cyclic sum and* $\mathcal{A}_{\{i,j\}}$ *represents an alternate sum.*

Proof: Let $(X_A) = \left(\delta/\delta t, \delta/\delta x^i, \partial/\partial y^i_1 \right)$ be the adapted basis of vector fields produced by the nonlinear connection (3.1). Let ∇ be the h-normal Γ-linear connection of Cartan type given by the components (3.4). For the linear connection ∇, the following general local Bianchi identities are true (see [55] or [63]):

$$\sum_{\{A,B,C\}} \left\{ \mathbf{R}^F_{ABC} - \mathbf{T}^F_{AB:C} - \mathbf{T}^G_{AB} \mathbf{T}^F_{CG} \right\} = 0,$$

$$\sum_{\{A,B,C\}} \left\{ \mathbf{R}^F_{DAB:C} + \mathbf{T}^G_{AB} \mathbf{R}^F_{DCG} \right\} = 0,$$

where $\mathbf{R}(X_A, X_B)X_C = \mathbf{R}^D_{CBA}X_D$, $\mathbf{T}(X_A, X_B) = \mathbf{T}^D_{BA}X_D$ and "$_{:A}$" represents one from the local covariant derivatives "$_{/1}$," "$_{|i}$," or "$|^{(1)}_{(i)}$." Obviously, the components $\mathbf{T}^C_{AB}$ and $\mathbf{R}^D_{ABC}$ are the adapted components of the torsion and curvature tensors associated to the h-normal Γ-linear connection of Cartan type $\nabla\Gamma$ given by (3.4). These components are expressed in Tables 3.5 and 3.6. Then, replacing $A, B, C, \ldots$ with indices of type

$$\left\{ 1, i, \,^{(1)}_{(i)} \right\},$$

by laborious local computations, we obtain the required Bianchi identities.

Remark 68. The above *eleven "star"-Bianchi identities* are exactly those eleven Bianchi identities that characterize the canonical metrical Cartan connection of a Finsler space (see [56], p. 48).

Remark 69. The importance of preceding Bianchi identities for an h-normal Γ-linear connection of Cartan type ∇ on the 1-jet space $J^1(\mathbb{R}, M)$ comes from their use in the local description of the *geometrical Maxwell equations* that characterize the jet single-time electromagnetic field $\mathbb{F}$ from the background of our relativistic time-dependent Lagrange geometry.

CHAPTER 4

THE JET RIEMANN-LAGRANGE GEOMETRY OF THE RELATIVISTIC TIME-DEPENDENT LAGRANGE SPACES

This Chapter develops the Riemann-Lagrange geometry (in the sense of *canonical nonlinear connection* Γ, *Cartan canonical* Γ-*linear connection*, together with its *d-torsions* and *d-curvatures*) produced only by a given *Kronecker h-regular Lagrangian function* on the 1-jet space $J^1(\mathbb{R}, M)$. This naturally extends the Miron-Anastasiei geometrical framework from [55].

At the same time, we construct a geometrical jet relativistic time-dependent Lagrangian theory of physical fields, which blends the electromagnetic and the gravitational theories. The developed generalized-theoretical field theory is governed by natural geometrical Maxwell and Einstein equations, which generalize the classical ones.

In order to have a clear exposition of our ideas, we point out that, in our geometrical development, we use the following two distinct notions:[1]

1. *time-dependent Lagrangian function* = a smooth function L on $J^1(\mathbb{R}, M)$;

---

[1] We note that if L is a Lagrangian function on $J^1(\mathbb{R}, M)$, then $\mathcal{L} = L\sqrt{h_{11}}$, where h_{11} is a Riemannian metric on $\mathbb{R}$, represents a Lagrangian on $J^1(\mathbb{R}, M)$.

Jet Single-Time Lagrange Geometry and Its Applications
1-st Edition. By Vladimir Balan and Mircea Neagu.
© 2011 John Wiley & Sons, Inc. Published 2011 John Wiley & Sons, Inc.

2. *time-dependent Lagrangian* (in P.J. Olver's terminology [77]) = a smooth local function $\mathcal{L}$ on $J^1(\mathbb{R}, M)$, which transforms by the rule $\widetilde{\mathcal{L}} = \mathcal{L} \left| dt/d\widetilde{t} \right|$.

We also underline that we generally use the terminology "time-dependent" instead of the equivalent "non-autonomous" or "rheonomic" concepts used by other authors.

4.1 RELATIVISTIC TIME-DEPENDENT LAGRANGE SPACES

To develop our jet time-dependent Lagrange geometry, we start the study considering $L : E \rightarrow \mathbb{R}$ a smooth Lagrangian function on $E = J^1(\mathbb{R}, M)$, which is locally expressed by $E \ni (t, x^i, y_1^i) \rightarrow L(t, x^i, y_1^i) \in \mathbb{R}$. The *vertical fundamental metrical d-tensor* of L is defined by

$$G_{(i)(j)}^{(1)(1)} = \frac{1}{2} \frac{\partial^2 L}{\partial y_1^i \partial y_1^j}. \tag{4.1}$$

Let $h = (h_{11}(t))$ be a Riemannian metric on the time manifold $\mathbb{R}$.

Definition 70. A jet Lagrangian function $L : E \rightarrow \mathbb{R}$, whose vertical fundamental metrical d-tensor (4.1) is of the form

$$G_{(i)(j)}^{(1)(1)}(t, x^k, y_1^k) := \frac{1}{2} \frac{\partial^2 L}{\partial y_1^i \partial y_1^j} = h^{11}(t) g_{ij}(t, x^k, y_1^k),$$

where $h^{11} = 1/h_{11} > 0$ and $g_{ij}(t, x^k, y_1^k)$ is a distinguished tensor on E, symmetric, of rank $n = \dim M$, and having a constant signature, is called a *Kronecker h-regular Lagrangian function with respect to the temporal Riemannian metric* $h = (h_{11})$.

In this context, we introduce the following concept:

Definition 71. A pair $RRL_1^n = (J^1(\mathbb{R}, M), L)$, where $n = \dim M$ and L is a Kronecker h-regular Lagrangian function, is called a *relativistic time-dependent (non-autonomous or rheonomic) Lagrange space* or a *jet single-time Lagrange space*.

■ **EXAMPLE 4.1**

Suppose that the spatial manifold M is endowed with a semi-Riemannian metric $\varphi = (\varphi_{ij}(x))$. Then, the time-dependent Lagrangian function

$$BS : J^1(\mathbb{R}, M) \rightarrow \mathbb{R},$$

defined by

$$BS = h^{11}(t) \varphi_{ij}(x) y_1^i y_1^j,$$

is a Kronecker h-regular Lagrangian function. Consequently, the pair

$$\mathcal{BSL}_1^n = (J^1(\mathbb{R}, M), BS)$$

is a relativistic time-dependent Lagrange space. It is important to note that the Lagrangian

$$\mathcal{B}\mathcal{S} = BS\sqrt{h_{11}} = h^{11}(t)\varphi_{ij}(x)y_1^i y_1^j\sqrt{h_{11}}$$

is exactly the energy Lagrangian whose extremals are the harmonic maps between the manifolds $(\mathbb{R}, h)$ and (M, φ) (see Eells and Lemaire [33]). At the same time, the Lagrangian $\mathcal{B}\mathcal{S}$ is a basic object in the physical theory of *bosonic strings*.

■ **EXAMPLE 4.2**

Let us consider the Kronecker h-regular Lagrangian function

$$NED : J^1(\mathbb{R}, M) \to \mathbb{R},$$

given by

$$NED = h^{11}(t)g_{ij}(t, x^k)y_1^i y_1^j + U_{(i)}^{(1)}(t, x^k)y_1^i + \Phi(t, x^k),$$

where $g_{ij}(t, x^k)$ is a d-tensor field on $E = J^1(\mathbb{R}, M)$, symmetric, of rank n, and having a constant signature, $U_{(i)}^{(1)}(t, x^k)$ are the local components of a d-tensor on the 1-jet space $E = J^1(\mathbb{R}, M)$, and $\Phi(t, x^k)$ is a smooth function on the product manifold $\mathbb{R} \times M$. Then, the pair

$$\mathcal{N}\mathcal{E}\mathcal{D}L_1^n = (J^1(\mathbb{R}, M), NED)$$

is a relativistic time-dependent Lagrange space, which is called the *non-autonomous jet single-time Lagrange space of electrodynamics*. Physically speaking, the dynamical character of the spatial gravitational potentials $g_{ij}(t, x^k)$ (i.e., the dependence of the temporal coordinate t) forced us to use the term *non-autonomous* in the preceding definition.

4.2   THE CANONICAL NONLINEAR CONNECTION

Let us consider a relativistic time-dependent Lagrange space $RRL_1^n = (J^1(\mathbb{R}, M), L)$, where L is a Kronecker h-regular Lagrangian function. Let $[a, b] \subset \mathbb{R}$ be a compact interval in the time manifold $\mathbb{R}$. In this context, we can define the *energy action functional* of the space RRL_1^n, setting

$$\mathbb{E} : C^\infty([a, b], M) \to \mathbb{R}, \quad \mathbb{E}(c) = \int_a^b L(t, x^i, y_1^i)\sqrt{h_{11}(t)}dt,$$

where the local expression of the smooth curve c on M is $t \to (x^i(t))$, and we have $y_1^i = dx^i/dt$.

The extremals of the energy action functional $\mathbb{E}$ verifies the Euler-Lagrange equations

$$2G^{(1)(1)}_{(i)(j)}\frac{d^2x^j}{dt^2} + \frac{\partial^2 L}{\partial x^j \partial y_1^i}\frac{dx^j}{dt} - \frac{\partial L}{\partial x^i} + \frac{\partial^2 L}{\partial t \partial y_1^i} + \frac{\partial L}{\partial y_1^i}\kappa_{11}^1 = 0, \quad \forall\, i = \overline{1,n}, \quad (4.2)$$

where

$$\kappa_{11}^1 = \frac{h^{11}}{2}\frac{dh_{11}}{dt}$$

is the Christoffel symbol of the Riemannian metric h_{11}.

Taking into account the Kronecker h-regularity of the Lagrangian function L, we can rearrange the Euler-Lagrange equations (4.2) of the Lagrangian $\mathcal{L} = L\sqrt{h_{11}}$ in the *Poisson form* (see also [73])

$$\Delta_h x^k + 2\mathcal{G}^k(t, x^m, y_1^m) = 0, \quad \forall\, k = \overline{1,n}, \quad (4.3)$$

where

$$\Delta_h x^k = h^{11}\left\{\frac{d^2x^k}{dt^2} - \kappa_{11}^1\frac{dx^k}{dt}\right\}, \quad y_1^m = \frac{dx^m}{dt},$$

$$2\mathcal{G}^k = \frac{g^{ki}}{2}\left\{\frac{\partial^2 L}{\partial x^j \partial y_1^i}y_1^j - \frac{\partial L}{\partial x^i} + \frac{\partial^2 L}{\partial t \partial y_1^i} + \frac{\partial L}{\partial y_1^i}\kappa_{11}^1 + 2h^{11}\kappa_{11}^1 g_{ij}y_1^j\right\}. \tag{4.4}$$

Theorem 72. *The geometrical object*

$$G = \left(G^{(r)}_{(1)1}\right),$$

where $G^{(r)}_{(1)1} = h_{11}\mathcal{G}^r$, is a spatial semispray on the 1-jet space E.

Proof: By a direct calculation, we deduce that the jet local geometrical entities

$$2\mathcal{S}^k = \frac{g^{ki}}{2}\left\{\frac{\partial^2 L}{\partial x^j \partial y_1^i}y_1^j - \frac{\partial L}{\partial x^i}\right\},$$

$$2\mathcal{H}^k = \frac{g^{ki}}{2}\left\{\frac{\partial^2 L}{\partial t \partial y_1^i} + \frac{\partial L}{\partial y_1^i}\kappa_{11}^1\right\}, \tag{4.5}$$

$$2\mathcal{J}^k = h^{11}\kappa_{11}^1 y_1^k$$

verify the following transformation rules:

$$2\mathcal{S}^p = 2\widetilde{\mathcal{S}}^q\frac{\partial x^p}{\partial \widetilde{x}^q} + h^{11}\frac{\partial x^p}{\partial \widetilde{x}^l}\frac{d\widetilde{t}}{dt}\frac{\partial \widetilde{y}_1^l}{\partial x^j}y_1^j,$$

$$2\mathcal{H}^p = 2\widetilde{\mathcal{H}}^q\frac{\partial x^p}{\partial \widetilde{x}^q} + h^{11}\frac{\partial x^p}{\partial \widetilde{x}^l}\frac{d\widetilde{t}}{dt}\frac{\partial \widetilde{y}_1^l}{\partial t},$$

$$2\mathcal{J}^p = 2\widetilde{\mathcal{J}}^q\frac{\partial x^p}{\partial \widetilde{x}^q} - h^{11}\frac{\partial x^p}{\partial \widetilde{x}^l}\frac{d\widetilde{t}}{dt}\frac{\partial \widetilde{y}_1^l}{\partial t}.$$

Consequently, the local entities $2\mathcal{G}^p = 2\mathcal{S}^p + 2\mathcal{H}^p + 2\mathcal{J}^p$ modify by the rules

$$2\widetilde{\mathcal{G}}^r = 2\mathcal{G}^p \frac{\partial \widetilde{x}^r}{\partial x^p} - \widetilde{h}^{11} \frac{\partial x^p}{\partial \widetilde{x}^j} \frac{\partial \widetilde{y}_1^r}{\partial x^p} \widetilde{y}_1^j. \qquad (4.6)$$

Multiplying now the relation (4.6) by $\widetilde{h}_{11}$, we find the transformation rules (1.10), which characterize a spatial semispray on $J^1(\mathbb{R}, M)$.

Taking into account the harmonic curve equations (1.12) of a time-dependent semispray on $J^1(\mathbb{R}, M)$, we can give the following natural geometrical interpretation to the Euler-Lagrange equations (4.2) and to their Poisson form (4.3), produced by the Lagrangian $\mathcal{L} = L\sqrt{h_{11}}$:

Theorem 73. *The extremals of the energy action functional* $\mathbb{E}$ *(produced by the Kronecker h-regular Lagrangian function L of the space* RRL_1^n*) are harmonic curves of the relativistic time-dependent semispray* $S_L = (H, G)$*, which is defined by the temporal components*

$$H^{(i)}_{(1)1} = -\frac{1}{2}\kappa^1_{11} y_1^i \qquad (4.7)$$

and the spatial components

$$G^{(i)}_{(1)1} = \frac{h_{11}g^{ik}}{4} \left[\frac{\partial^2 L}{\partial x^j \partial y_1^k} y_1^j - \frac{\partial L}{\partial x^k} + \frac{\partial^2 L}{\partial t \partial y_1^k} + \frac{\partial L}{\partial y_1^k} \kappa^1_{11} + 2h^{11}\kappa^1_{11}g_{kl}y_1^l \right]. \qquad (4.8)$$

In other words, these extremals verify the equations of the **harmonic curves**

$$h^{11} \left\{ \frac{d^2 x^i}{dt^2} + 2H^{(i)}_{(1)1} + 2G^{(i)}_{(1)1} \right\} = 0, \quad \forall\, i = \overline{1, n}, \ n = \dim M.$$

Definition 74. The relativistic time-dependent semispray

$$S_L = \left(H^{(i)}_{(1)1}, G^{(i)}_{(1)1} \right),$$

defined by the components (4.7) and (4.8), is called the *canonical time-dependent semispray of the jet single-time Lagrange space* RRL_1^n.

Using now Proposition 26, we can introduce the following concept:

Definition 75. The nonlinear connection

$$\Gamma_L := \Gamma = \left(M^{(i)}_{(1)1}, N^{(i)}_{(1)j} \right),$$

given by

$$M^{(i)}_{(1)1} = 2H^{(i)}_{(1)1} = -\kappa^1_{11} y_1^i, \qquad N^{(i)}_{(1)j} = \frac{\partial G^{(i)}_{(1)1}}{\partial y_1^j}, \qquad (4.9)$$

where $H^{(i)}_{(1)1}$ and $G^{(i)}_{(1)1}$ are defined by formulas (4.7) and (4.8), is called the *canonical nonlinear connection of the relativistic time-dependent Lagrange space* RRL_1^n.

4.3 THE CARTAN CANONICAL METRICAL LINEAR CONNECTION

The main result of this Chapter is the Theorem of Existence and Uniqueness of the *Cartan canonical h-normal Γ-linear connection $C\Gamma$*, which allows the subsequent development of our *jet single-time Lagrangian theory of physical fields* on a relativistic time-dependent Lagrange space RRL_1^n.

Theorem 76 (Cartan canonical linear connection). *On the relativistic time-dependent Lagrange space $RRL_1^n = (J^1(\mathbb{R}, M), L)$ endowed with its canonical nonlinear connection Γ [see formulas (4.9)] there exists an unique h-normal Γ-linear connection*

$$C\Gamma = \left(\kappa_{11}^1,\ G_{j1}^k,\ L_{jk}^i,\ C_{j(k)}^{i(1)} \right)$$

having the metrical properties

1. $g_{ij|k} = 0, \quad g_{ij}\big|_{(k)}^{(1)} = 0,$

2. $G_{j1}^k = \dfrac{g^{km}}{2}\dfrac{\delta g_{mj}}{\delta t}, \quad L_{jk}^i = L_{kj}^i, \quad C_{j(k)}^{i(1)} = C_{k(j)}^{i(1)}.$

Proof: Let us consider that

$$C\Gamma = \left(\overline{G}_{11}^1,\ G_{j1}^k,\ L_{jk}^i,\ C_{j(k)}^{i(1)} \right)$$

is an h-normal Γ-linear connection whose coefficients are defined by the relations

$$\overline{G}_{11}^1 = \kappa_{11}^1, \qquad G_{j1}^k = \frac{g^{km}}{2}\frac{\delta g_{mj}}{\delta t},$$

$$L_{jk}^i = \frac{g^{im}}{2}\left(\frac{\delta g_{jm}}{\delta x^k} + \frac{\delta g_{km}}{\delta x^j} - \frac{\delta g_{jk}}{\delta x^m} \right),$$

$$C_{j(k)}^{i(1)} = \frac{g^{im}}{2}\left(\frac{\partial g_{jm}}{\partial y_1^k} + \frac{\partial g_{km}}{\partial y_1^j} - \frac{\partial g_{jk}}{\partial y_1^m} \right).$$

Then, by direct computations, one verifies that $C\Gamma$ satisfies conditions 1 and 2.

Conversely, let us consider that the h-normal Γ-linear connection

$$\widetilde{C}\Gamma = \left(\widetilde{\overline{G}}_{11}^1,\ \widetilde{G}_{j1}^k,\ \widetilde{L}_{jk}^i,\ \widetilde{C}_{j(k)}^{i(1)} \right)$$

satisfies conditions 1 and 2. It follows directly that we have

$$\widetilde{\overline{G}}_{11}^1 = \kappa_{11}^1, \qquad \widetilde{G}_{j1}^k = \frac{g^{km}}{2}\frac{\delta g_{mj}}{\delta t}.$$

The condition $g_{ij|k} = 0$ is equivalent to

$$\frac{\delta g_{ij}}{\delta x^k} = g_{mj}\widetilde{L}_{ik}^m + g_{im}\widetilde{L}_{jk}^m.$$

Applying a Christoffel process to the indices $\{i, j, k\}$ and using the symmetry relations $\widetilde{L}^i_{jk} = \widetilde{L}^i_{kj}$, we find

$$\widetilde{L}^i_{jk} = \frac{g^{im}}{2} \left(\frac{\delta g_{jm}}{\delta x^k} + \frac{\delta g_{km}}{\delta x^j} - \frac{\delta g_{jk}}{\delta x^m} \right).$$

By analogy, using the relations $\widetilde{C}^{i(1)}_{j(k)} = \widetilde{C}^{i(1)}_{k(j)}$ and $g_{ij}|^{(1)}_{(k)} = 0$, a Christoffel process applied to the indices $\{i, j, k\}$ leads to

$$\widetilde{C}^{i(1)}_{j(k)} = \frac{g^{im}}{2} \left(\frac{\partial g_{jm}}{\partial y^k_1} + \frac{\partial g_{km}}{\partial y^j_1} - \frac{\partial g_{jk}}{\partial y^m_1} \right).$$

In conclusion, the existence and the uniqueness of the Cartan canonical linear connection $C\Gamma$ is clear.

Remark 77. Replacing the canonical nonlinear connection Γ by an arbitrary one, the preceding Theorem still holds true.

Remark 78. As a rule, the Cartan canonical linear connection of the relativistic time-dependent Lagrange space RRL^n_1 verifies also the metrical properties

$$h_{11/1} = h_{11|k} = h_{11}|^{(1)}_{(k)} = 0, \qquad g_{ij/1} = 0.$$

Remark 79. In the particular case $g_{ij}(t, x^k, y^k_1) = g_{ij}(x^k) := \varphi_{ij}(x^k)$, the coefficients of the Cartan linear connection $C\Gamma$ are the same with those of the Berwald linear connection $B\overset{\circ}{\Gamma}$, namely,

$$C\Gamma = B\overset{\circ}{\Gamma} = \left(\kappa^1_{11}, \ 0, \ \gamma^i_{jk}, \ 0 \right).$$

However, it is important to note that the Cartan connection is a Γ-linear connection, where Γ is the canonical nonlinear connection of the space RRL^n_1, while the Berwald connection is a $\overset{\circ}{\Gamma}$-linear connection, where $\overset{\circ}{\Gamma}$ is the canonical nonlinear connection produced by the pair of metrics $(h_{11}(t), \varphi_{ij}(x))$. Consequently, the Cartan connection and the Berwald connection are distinct ones in this case, even if they have the same local adapted coefficients.

Remark 80. The torsion $\mathbf{T}$ of the Cartan canonical connection of the relativistic time-dependent Lagrange space RRL^n_1 is determined only by *six* effective local components, because the properties of the Cartan canonical connection imply the equalities (see Theorem 55)

$$T^r_{ij} = 0, \qquad S^{(r)(1)(1)}_{(1)(i)(j)} = 0.$$

At the same time, it is important to note that the number of the curvature local d-tensors of the Cartan canonical connection does not reduce. *Five* d-curvatures remain, their formulas being given by Theorem 58.

Definition 81. The torsion and curvature d-tensors of the Cartan canonical connection of the space RRL^n_1 (given by Theorem 55 and Theorem 58) are called the *torsions* and *curvatures of the relativistic time-dependent Lagrange space* RRL^n_1.

4.4 RELATIVISTIC TIME-DEPENDENT LAGRANGIAN ELECTROMAGNETISM

4.4.1 The jet single-time electromagnetic field

Let us consider that $RRL_1^n = (J^1(\mathbb{R}, M), L)$ is a relativistic time-dependent Lagrange space and

$$\Gamma = \left(M_{(1)1}^{(i)}, N_{(1)j}^{(i)} \right)$$

is its canonical nonlinear connection. At the same time, let

$$C\Gamma = \left(\kappa_{11}^1, \ G_{j1}^k, \ L_{jk}^i, \ C_{j(k)}^{i(1)} \right)$$

be the Cartan canonical connection of the space RRL_1^n. In this context, using the canonical Liouville d-tensor field

$$\mathbf{C} := \mathbf{C}_{(1)}^{(i)} \frac{\partial}{\partial y_1^i} = y_1^i \frac{\partial}{\partial y_1^i},$$

we can introduce the following nonmetrical deflection d-tensors:

$$\overline{D}_{(1)1}^{(i)} = \mathbf{C}_{(1)/1}^{(i)} = y_{1/1}^i, \quad D_{(1)j}^{(i)} = \mathbf{C}_{(1)|j}^{(i)} = y_{1|j}^i, \quad d_{(1)(j)}^{(i)(1)} = \mathbf{C}_{(1)}^{(i)}|_{(j)}^{(1)} = y_1^i|_{(j)}^{(1)},$$

where "$/1$," "$|_j$," and "$|_{(j)}^{(1)}$" are the local covariant derivatives induced by the Cartan canonical connection $C\Gamma$. It follows that, by direct calculations, we find the following:

Proposition 82. *The nonmetrical deflection d-tensors of the relativistic time-dependent Lagrange space RRL_1^n have the expressions*

$$\overline{D}_{(1)1}^{(i)} = \frac{g^{il}}{2} \frac{\delta g_{lm}}{\delta t} y_1^m, \quad D_{(1)j}^{(i)} = -N_{(1)j}^{(i)} + L_{jm}^i y_1^m, \quad d_{(1)(j)}^{(i)(1)} = \delta_j^i + C_{m(j)}^{i(1)} y_1^m.$$

In the sequel, using the vertical fundamental metrical d-tensor

$$G_{(i)(j)}^{(1)(1)} = h^{11} g_{ij},$$

we construct the *metrical deflection d-tensors* of the relativistic time-dependent Lagrange space RRL_1^n:

$$\overline{D}_{(i)1}^{(1)} := G_{(i)(r)}^{(1)(1)} \overline{D}_{(1)1}^{(r)} = \frac{h^{11}}{2} \frac{\delta g_{im}}{\delta t} y_1^m,$$

$$D_{(i)j}^{(1)} := G_{(i)(r)}^{(1)(1)} D_{(1)j}^{(r)} = h^{11} g_{ir} \left[-N_{(1)j}^{(r)} + L_{jm}^r y_1^m \right],$$

$$d_{(i)(j)}^{(1)(1)} := G_{(i)(r)}^{(1)(1)} d_{(1)(j)}^{(r)(1)} = h^{11} \left[g_{ij} + g_{ir} C_{m(j)}^{r(1)} y_1^m \right].$$

Definition 83. The distinguished 2-form on $E = J^1(\mathbb{R}, M)$, given by

$$\mathbb{F} = F_{(i)j}^{(1)} \delta y_1^i \wedge dx^j + f_{(i)(j)}^{(1)(1)} \delta y_1^i \wedge \delta y_1^j,$$

where

$$F_{(i)j}^{(1)} = \frac{1}{2}\left[D_{(i)j}^{(1)} - D_{(j)i}^{(1)}\right], \qquad f_{(i)(j)}^{(1)(1)} = \frac{1}{2}\left[d_{(i)(j)}^{(1)(1)} - d_{(j)(i)}^{(1)(1)}\right],$$

is called the *jet single-time electromagnetic field* of the relativistic time-dependent Lagrange space RRL_1^n.

Using the above Definition, by a direct calculation, we obtain the following:

Proposition 84. *The expressions of the electromagnetic components are*

$$F_{(i)j}^{(1)} = \frac{h^{11}}{2}\left[g_{jr}N_{(1)i}^{(r)} - g_{ir}N_{(1)j}^{(r)} + \left(g_{ir}L_{jm}^r - g_{jr}L_{im}^r\right)y_1^m\right],$$

$$f_{(i)(j)}^{(1)(1)} = 0.$$

(4.10)

Remark 85. The above Proposition shows us that the jet single-time electromagnetic field of the relativistic time-dependent Lagrange space RRL_1^n has, in fact, the following simpler form:

$$\mathbb{F} = F_{(i)j}^{(1)}\delta y_1^i \wedge dx^j,$$

where the electromagnetic components $F_{(i)j}^{(1)}$ are given by (4.10).

4.4.2   Geometrical Maxwell equations

The main result of the relativistic time-dependent Lagrangian electromagnetism is the following:

Theorem 86. *The electromagnetic components $F_{(i)j}^{(1)}$ of the relativistic time-dependent Lagrange space RRL_1^n are governed by the following* **geometrical Maxwell equations**:

$$\begin{cases} F_{(i)k/1}^{(1)} = \frac{1}{2}\mathcal{A}_{\{i,k\}}\left\{\overline{D}_{(i)1|k}^{(1)} + D_{(i)m}^{(1)}T_{1k}^m + d_{(i)(m)}^{(1)(1)}R_{(1)1k}^{(m)} - \left[T_{1i|k}^p + C_{k(m)}^{p(1)}R_{(1)1i}^{(m)}\right]y_p^1\right\}, \\[2mm] \sum_{\{i,j,k\}}F_{(i)j|k}^{(1)} = -\sum_{\{i,j,k\}}C_{(i)(m)(r)}^{(1)(1)(1)}R_{(1)jk}^{(r)}y_1^m, \\[2mm] \sum_{\{i,j,k\}}F_{(i)j}^{(1)}|_{(k)}^{(1)} = 0, \end{cases}$$

where $\mathcal{A}_{\{i,k\}}$ means an alternate sum, $\sum_{\{i,j,k\}}$ means a cyclic sum, and

$$y_p^1 = h^{11}g_{pq}y_1^q, \qquad C_{(i)(m)(r)}^{(1)(1)(1)} = h^{11}g_{mq}C_{i(r)}^{q(1)} = \frac{1}{4}\frac{\partial^3 L}{\partial y_1^i \partial y_1^m \partial y_1^1}.$$

Proof: First, we point out that the Ricci identities applied to the spatial metrical d-tensor g_{ij} lead us to the following curvature d-tensor identities:

$$R_{mi1k} + R_{im1k} = 0, \quad R_{mijk} + R_{imjk} = 0, \quad P_{mij(k)}^{(1)} + P_{imj(k)}^{(1)} = 0,$$

where $R_{mi1k} = g_{ip}R^p_{m1k}$, $R_{mijk} = g_{ip}R^p_{mjk}$, and $P_{mij(k)}^{\ \ (1)} = g_{ip}P^{p\ (1)}_{mj(k)}$.

Now, let us consider the following nonmetrical deflection d-tensor identities attached to the Cartan canonical connection:

(d_1) $\bar{D}^{(p)}_{(1)1|k} - D^{(p)}_{(1)k/1} = y^m_1 R^p_{m1k} - D^{(p)}_{(1)m}T^m_{1k} - d^{(p)(1)}_{(1)(m)}R^{(m)}_{(1)1k}$,

(d_2) $D^{(p)}_{(1)j|k} - D^{(p)}_{(1)k|j} = y^m_1 R^p_{mjk} - d^{(p)(1)}_{(1)(m)}R^{(m)}_{(1)jk}$,

(d_3) $D^{(p)}_{(1)j}|^{(1)}_{(k)} - d^{(p)(1)}_{(1)(k)|j} = y^m_1 P^{p\ (1)}_{mj(k)} - D^{(p)}_{(1)m}C^{m(1)}_{j(k)} - d^{(p)(1)}_{(1)(m)}P^{(m)\ (1)}_{(1)j(k)}$.

Contracting these nonmetrical deflection d-tensor identities by $G^{(1)(1)}_{(i)(p)} = h^{11}g_{ip}$, we obtain the following metrical deflection d-tensor identities:

(m_1) $\bar{D}^{(1)}_{(i)1|k} - D^{(1)}_{(i)k/1} = h^{11}y^m_1 R_{mi1k} - D^{(1)}_{(i)m}T^m_{1k} - d^{(1)(1)}_{(i)(m)}R^{(m)}_{(1)1k}$,

(m_2) $D^{(1)}_{(i)j|k} - D^{(1)}_{(i)k|j} = h^{11}y^m_1 R_{mijk} - d^{(1)(1)}_{(i)(m)}R^{(m)}_{(1)jk}$,

(m_3) $D^{(1)}_{(i)j}|^{(1)}_{(k)} - d^{(1)(1)}_{(i)(k)|j} = h^{11}y^m_1 P^{\ \ (1)}_{mij(k)} - D^{(1)}_{(i)m}C^{m(1)}_{j(k)} - d^{(1)(1)}_{(i)(m)}P^{(m)\ (1)}_{(1)j(k)}$.

At the same time, we recall that the following Bianchi identities are true:

(b_1) $\mathcal{A}_{\{i,k\}}\left\{R^l_{i1k} + T^l_{1i|k} + C^{l(1)}_{k(r)}R^{(r)}_{(1)1i}\right\} = 0$,

(b_2) $\sum_{\{i,j,k\}}\left\{R^l_{ijk} - C^{l(1)}_{i(r)}R^{(r)}_{(1)jk}\right\} = 0$,

(b_3) $\mathcal{A}_{\{i,j\}}\left\{P^{l\ (1)}_{ij(k)} + C^{l(1)}_{i(k)|j} + C^{l(1)}_{j(r)}P^{(r)\ (1)}_{(1)i(k)}\right\} = 0$.

In order to obtain the first Maxwell equation, we do an alternate sum $\mathcal{A}_{\{i,k\}}$ in (m_1). Using now the Bianchi identity (b_1), we obtain what we are looking for.

Doing a cyclic sum by the indices $\{i, j, k\}$ in (m_2) and using the Bianchi identity (b_2), it follows the second Maxwell equation. Note that, in these calculations, we also use the identity (which is proved by enough complicated direct computations)

$$\sum_{\{i,j,k\}}\left\{g_{im}R^{(m)}_{(1)jk}\right\} = 0.$$

Doing an alternate sum $\mathcal{A}_{\{i,j\}}$ in (m_3) and combining with the Bianchi identity (b_3), we obtain a new identity. A cyclic sum by the indices $\{i, j, k\}$ applied to this last identity implies the third Maxwell equation. Note that we also use in these computations the symmetry relations

$$d^{(1)(1)}_{(j)(k)} = d^{(1)(1)}_{(k)(j)}, \qquad C^{m(1)}_{j(k)} = C^{m(1)}_{k(j)}, \qquad P^{(m)\ (1)}_{(1)j(k)} = P^{(m)\ (1)}_{(1)k(j)}.$$

4.5   JET RELATIVISTIC TIME-DEPENDENT LAGRANGIAN GRAVITATIONAL THEORY

4.5.1   The jet single-time gravitational field

Let us consider a relativistic time-dependent Lagrange space $RRL^n_1 = (J^1(\mathbb{R}, M), L)$, whose Kronecker h-regular Lagrangian function L produces the vertical fundamental metrical d-tensor

$$G^{(1)(1)}_{(i)(j)} = \frac{1}{2}\frac{\partial^2 L}{\partial y_1^i \partial y_1^j} = h^{11}(t)g_{ij}(t, x^k, y_1^k).$$

Let us consider that

$$\Gamma = \left(M^{(i)}_{(1)1}, N^{(i)}_{(1)j}\right)$$

is the canonical nonlinear connection of the space RRL_1^n and let

$$C\Gamma = \left(\kappa_{11}^1,\ G_{j1}^k,\ L_{jk}^i,\ C_{j(k)}^{i(1)}\right)$$

be the Cartan canonical connection of the space RRL_1^n.

In order to develop on $J^1(\mathbb{R}, M)$ the relativistic time-dependent Lagrangian theory of gravitational field, we introduce the following geometrical concept with physical meaning:

Definition 87. The adapted metrical d-tensor $\mathbb{G}$ on the 1-jet space $J^1(\mathbb{R}, M)$, locally expressed by

$$\mathbb{G} = h_{11}dt \otimes dt + g_{ij}dx^i \otimes dx^j + h^{11}g_{ij}\delta y_1^i \otimes \delta y_1^j,$$

is called the *jet single-time gravitational field* of the relativistic time-dependent Lagrange space RRL_1^n.

4.5.2   Geometrical Einstein equations and conservation laws

We postulate that the *geometrical Einstein equations*, which govern the jet single-time gravitational field $\mathbb{G}$ of the space RRL_1^n, are the geometrical Einstein equations attached to the Cartan canonical connection $C\Gamma$, namely,

$$\text{Ric}(C\Gamma) - \frac{\text{Sc}(C\Gamma)}{2}\mathbb{G} = \mathcal{K}\mathcal{T}, \tag{4.11}$$

where $\text{Ric}(C\Gamma)$ represents the Ricci d-tensor of the Cartan connection, $\text{Sc}(C\Gamma)$ is its scalar curvature, $\mathcal{K}$ is the Einstein constant, and $\mathcal{T}$ is a given intrinsic d-tensor of matter, which is called the *stress-energy d-tensor*.

In this context, working with the adapted basis of vector fields

$$\{X_A\} = \left\{\frac{\delta}{\delta t},\ \frac{\delta}{\delta x^i},\ \frac{\partial}{\partial y_1^i}\right\}, \tag{4.12}$$

where A (or any other capital Latin letter, like B, C, D, ...) is an index of type "$_1$," "$_i$," or "$^{(1)}_{(i)}$," the curvature d-tensor of the Cartan connection has the folowing adapted local components:

$$\mathbf{R}(X_C, X_B)X_A = \mathbf{R}^D_{ABC}X_D.$$

Therefore, the Ricci d-tensor $\text{Ric}(C\Gamma)$ has the adapted components

$$\mathbf{R}_{AB} = \text{Ric}(X_A, X_B) = \text{Trace}\{Z \mapsto \mathbf{R}(Z, X_B)X_A\} = \mathbf{R}^M_{ABM},$$

and the scalar curvature takes the form $\mathrm{Sc}(C\Gamma) = \mathbf{G}^{AB}\mathbf{R}_{AB}$, where

$$\mathbf{G}^{AB} = \begin{cases} h_{11}, & \text{for } A = 1,\ B = 1, \\ g^{ij}, & \text{for } A = i,\ B = j, \\ h_{11}g^{ij}, & \text{for } A = \overset{(1)}{(i)},\ B = \overset{(1)}{(j)}, \\ 0, & \text{otherwise.} \end{cases} \tag{4.13}$$

Remark 88. Because of the form of the Bianchi identities that characterize the Cartan canonical connection $C\Gamma$ of the space RRL_1^n (see Theorem 67), the Ricci d-tensor $\mathrm{Ric}(C\Gamma)$ is not necessarily symmetric.

Taking into account the expressions of the local curvature d-tensors $\mathbf{R}^D_{ABC}$ of the Cartan connection $C\Gamma$ (see Theorem 58), together with the form (4.13) of the d-tensor $\mathbf{G}^{AB}$, we find the following:

Proposition 89. *The Ricci d-tensor* $\mathrm{Ric}(C\Gamma)$ *is determined by the following* **six** *effective adapted local components:*

$$R_{11} := \kappa_{11} = 0, \quad R_{i1} = R^m_{i1m}, \quad R_{ij} = R^m_{ijm}, \quad R^{(1)}_{i(j)} := P^{(1)}_{i(j)} = -P^{m\,(1)}_{im(j)},$$

$$R^{(1)}_{(i)1} := P^{(1)}_{(i)1} = P^{m\,(1)}_{i1(m)}, \quad R^{(1)}_{(i)j} := P^{(1)}_{(i)j} = P^{m\,(1)}_{ij(m)}, \quad R^{(1)(1)}_{(i)(j)} := S^{(1)(1)}_{(i)(j)} = S^{m(1)(1)}_{i(j)(m)},$$

where $\kappa_{11} = 0$ *is the classical Ricci tensor of the Riemannian metric* h_{11}.

Denoting $\kappa = h^{11}\kappa_{11} = 0$, $R = g^{ij}R_{ij}$, and $S = h_{11}g^{ij}S^{(1)(1)}_{(i)(j)}$, we obtain the following:

Proposition 90. *The scalar curvature of the Cartan canonical connection has the expression*

$$\mathrm{Sc}\,(C\Gamma) = \kappa + R + S = R + S.$$

Using the above results, we can establish the main result of the relativistic time-dependent Lagrangian theory of gravitational field:

Theorem 91. *The local* **geometrical Einstein equations**, *which govern the jet single-time gravitational field* $\mathbb{G}$ *of the space* RRL_1^n, *have the form*

$$\begin{cases} -\dfrac{R+S}{2}h_{11} = \mathcal{K}\mathcal{T}_{11}, \\[2mm] R_{ij} - \dfrac{R+S}{2}g_{ij} = \mathcal{K}\mathcal{T}_{ij}, \\[2mm] S^{(1)(1)}_{(i)(j)} - \dfrac{R+S}{2}h^{11}g_{ij} = \mathcal{K}\mathcal{T}^{(1)(1)}_{(i)(j)}, \end{cases} \tag{4.14}$$

$$\begin{cases} 0 = \mathcal{T}_{1i}, \quad R_{i1} = \mathcal{K}\mathcal{T}_{i1}, \quad P^{(1)}_{(i)1} = \mathcal{K}\mathcal{T}^{(1)}_{(i)1}, \\[2mm] 0 = \mathcal{T}^{(1)}_{1(i)} \quad P^{(1)}_{i(j)} = \mathcal{K}\mathcal{T}^{(1)}_{i(j)}, \quad P^{(1)}_{(i)j} = \mathcal{K}\mathcal{T}^{(1)}_{(i)j}, \end{cases} \tag{4.15}$$

where $\mathcal{T}_{AB}$, $A, B \in \left\{ 1, i, \begin{smallmatrix}(1)\\(i)\end{smallmatrix} \right\}$, *are the adapted components of the stress-energy d-tensor* $\mathcal{T}$.

Proof: We locally describe the global Einstein equations (4.11) in the adapted basis (4.12).

Remark 92. In order to have the compatibility of the Einstein equations (4.14) and (4.15), it is necessary as certain adapted local components of the stress-energy d-tensor $\mathcal{T}$ to be *a priori* equal to zero.

It is well known that, from a physical point of view, the stress-energy d-tensor of matter must verify the local *geometrical conservation laws*

$$\mathcal{T}^M_{A|M} = 0, \quad \forall \, A \in \left\{ 1, i, \begin{smallmatrix}(1)\\(i)\end{smallmatrix} \right\},$$

where $\mathcal{T}^B_A = \mathbf{G}^{BD}\mathcal{T}_{DA}$.

Theorem 93. *In the relativistic time-dependent Lagrangian geometry of the gravitational field the following* **geometrical conservation laws** *of the Einstein equations must be true:*

$$\begin{cases} \left[\dfrac{R + S}{2} \right]_{/1} = R^m_{1|m} + P^{(m)}_{(1)1}|^{(1)}_{(m)} \\[2ex] \left[R^m_i - \dfrac{R + S}{2}\delta^m_i \right]_{|m} = -P^{(m)}_{(1)i}|^{(1)}_{(m)} \\[2ex] \left[S^{(m)(1)}_{(1)(i)} - \dfrac{R + S}{2}\delta^m_i \right]|^{(1)}_{(m)} = -P^{m(1)}_{(i)|m}, \end{cases}$$

where $R^m_1 = g^{mr}R_{r1}, P^{(m)}_{(1)1} = h_{11}g^{mr}P^{(1)}_{(r)1}, R^m_i = g^{mr}R_{ri}, P^{(m)}_{(1)i} = h_{11}g^{mr}P^{(1)}_{(r)i},$
$S^{(m)(1)}_{(1)(i)} = h_{11}g^{mr}S^{(1)(1)}_{(r)(i)}, and\ P^{m(1)}_{(i)} = g^{mr}P^{(1)}_{r(i)}.$

CHAPTER 5

THE JET SINGLE-TIME ELECTRODYNAMICS

The aim of this Chapter is to develop the Riemann-Lagrange differential geometry (in the sense of Cartan linear connection, d-torsions, d-curvatures and geometrical Maxwell and Einstein equations) for the jet single-time Lagrange space of electrodynamics, which is governed by the quadratic Lagrangian function

$$L_{\text{jet}} : J^1(\mathbb{R}, M) \to \mathbb{R},$$

given by

$$L_{\text{jet}}(t, x^k, y_1^k) = h^{11}(t) \, \varphi_{ij}(x^k) \, y_1^i y_1^j + U_{(i)}^{(1)}(t, x^k) \, y_1^i + \Phi(t, x^k), \qquad (5.1)$$

where $h_{11}(t)$ [resp. $\varphi_{ij}(x^k)$] is a Riemannian (respectively, semi-Riemannian) metric on the temporal manifold $\mathbb{R}$ (resp. spatial manifold M), $U_{(i)}^{(1)}(t, x^k)$ are the local components of a d-tensor on the 1-jet space $J^1(\mathbb{R}, M)$ and $\Phi(t, x^k)$ is a smooth function on the product manifold $\mathbb{R} \times M$.

In order to justify our "electrodynamics" terminology, we recall that, in the study of classical electrodynamics, the Lagrangian function on the tangent bundle

$$L_{\text{autonomous}} : TM \to \mathbb{R},$$

Jet Single-Time Lagrange Geometry and Its Applications
1-st Edition. By Vladimir Balan and Mircea Neagu.
© 2011 John Wiley & Sons, Inc. Published 2011 John Wiley & Sons, Inc.

that governs the movement law of a particle of mass $m \neq 0$ and electric charge e, which is displaced concomitantly into an environment endowed both with a gravitational field and an electromagnetic one, is given by

$$L_{\text{autonomous}}(x, y) = mc\theta_{ij}(x)y^i y^j + \frac{2e}{m}A_i(x)y^i + \mathcal{F}(x). \qquad (5.2)$$

Here, the semi-Riemannian metric $\theta_{ij}(x)$ represents the *gravitational potentials* of the space of events M, $A_i(x)$ are the components of an 1-form on M representing the *electromagnetic potential*, $\mathcal{F}(x)$ is a smooth *potential function* on M, and c is the velocity of light in vacuum.

Remark 94. The classical Lagrange space $L^n = (M, L(x, y))$, where L is the Lagrangian function (5.2), is known in the literature of specialty as the *autonomous Lagrange space of electrodynamics*. A geometrical study of this space was completely done by Miron and Anastasiei in the book [55].

More generally, we point out that, in the study of classical *time-dependent* (*non-autonomous* or *rheonomic*) *electrodynamics*, a central role is played by the *non-autonomous Lagrangian function of electrodynamics*, $L_{\text{non-autonomous}} : \mathbb{R} \times TM \to \mathbb{R}$, expressed by

$$L_{\text{non-autonomous}}(t, x, y) = mc\theta_{ij}(x)y^i y^j + \frac{2e}{m}A_i(t, x)y^i + \mathcal{F}(t, x). \qquad (5.3)$$

Remark 95. The differential geometry produced by the *non-autonomous Lagrangian function of electrodynamics* (5.3) is also given in the book [55].

As a consequence, in our jet relativistic single-time approach of the electrodynamics, we consider that a central role is played by the following *jet relativistic time-dependent Lagrangian of electrodynamics*:

$$\mathcal{L} = \left(mch^{11}(t)\theta_{ij}(x)y_1^i y_1^j + \frac{2e}{m}A_{(i)}^{(1)}(t, x)y_1^i + \mathcal{F}(t, x) \right)\sqrt{h_{11}}, \qquad (5.4)$$

where $h_{11}(t)$ is a Riemannian metric on the time manifold $\mathbb{R}$, $A_{(i)}^{(1)}(t, x)$ is a d-tensor on $J^1(\mathbb{R}, M)$ and $\mathcal{F}(t, x)$ is a smooth function on $\mathbb{R} \times M$.

5.1   RIEMANN-LAGRANGE GEOMETRY ON THE JET SINGLE-TIME LAGRANGE SPACE OF ELECTRODYNAMICS $\mathcal{E}\mathcal{D}L_1^N$

In such a context, by a natural extension of Lagrangian functions (5.2), (5.3), and (5.4), we introduce

Definition 96. The pair $\mathcal{E}\mathcal{D}L_1^n = (J^1(\mathbb{R}, M), L_{\text{jet}})$, where L_{jet} is a quadratic single-time Lagrangian function of the form (5.1), is called the *autonomous jet single-time Lagrange space of electrodynamics*.

Remark 97. The non-dynamical character of the spatial metrical d-tensor $\varphi_{ij}(x^k)$ (i.e., the independence on the temporal coordinate t) determined us to use the term "autonomous" in the preceding definition.

In the sequel, we apply the general Riemann-Lagrange geometry of the relativistic time-dependent Lagrange spaces from the preceding chapters to the particular autonomous jet single-time Lagrange space of electrodynamics $\mathcal{E}\mathcal{D}L_1^n$.

To initiate the development of the Riemann-Lagrange geometry of the jet single-time Lagrange space $\mathcal{E}\mathcal{D}L_1^n$, endowed with the Lagrangian of electrodynamics

$$\mathcal{E}\mathcal{D} = L_{\text{jet}}\sqrt{h_{11}} = \left[h^{11}(t)\varphi_{ij}(x^k)y_1^i y_1^j + U_{(i)}^{(1)}(t,x^k)y_1^i + \Phi(t,x^k) \right]\sqrt{h_{11}},$$

let us consider the *energy action functional*, defined by

$$\mathbb{E}_{\mathcal{E}\mathcal{D}} : C^\infty([a,b], M) \to \mathbb{R}, \quad \mathbb{E}_{\mathcal{E}\mathcal{D}}(c) = \int_a^b \mathcal{E}\mathcal{D}dt = \int_a^b L_{\text{jet}}\sqrt{h_{11}}dt, \quad (5.1)$$

where the $[a,b] \subset \mathbb{R}$ is a compact interval, the local expression of the smooth curve c on M is $t \to (x^i(t))$, and $y_1^i = dx^i/dt$. In this context, a preceding general result implies (see also [65] and [66])

Theorem 98. *The extremals of the energy action functional $\mathbb{E}_{\mathcal{E}\mathcal{D}}$ (produced by the autonomous single-time Lagrangian of electrodynamics $\mathcal{E}\mathcal{D}$) are harmonic curves of the relativistic time-dependent semispray $\mathcal{S}_{\mathcal{E}\mathcal{D}} = (H, G)$, which is defined by the temporal components*

$$H_{(1)1}^{(i)} = -\frac{1}{2}\kappa_{11}^1 y_1^i$$

and the spatial components

$$G_{(1)1}^{(i)} = \frac{1}{2}\gamma_{pq}^i y_1^p y_1^q + \frac{h_{11}\varphi^{im}}{4}\left[U_{(m)p}^{(1)}y_1^p + \frac{\partial U_{(m)}^{(1)}}{\partial t} + U_{(m)}^{(1)}\kappa_{11}^1 - \frac{\partial \Phi}{\partial x^m} \right],$$

where $\gamma_{pq}^i(x)$ are the Christoffel symbols of the semi-Riemannian metric $\varphi_{ij}(x)$, the Christoffel symbol of the Riemannian metric $h_{11}(t)$ is

$$\kappa_{11}^1(t) = \frac{h^{11}}{2}\frac{dh_{11}}{dt},$$

where $h^{11} = 1/h_{11} > 0$, and we have the following:

$$U_{(m)p}^{(1)} := \frac{\partial U_{(m)}^{(1)}}{\partial x^p} - \frac{\partial U_{(p)}^{(1)}}{\partial x^m}.$$

*In other words, these extremals verify the equations of the **harmonic curves***

$$h^{11}\left\{ \frac{d^2x^i}{dt^2} + 2H_{(1)1}^{(i)} + 2G_{(1)1}^{(i)} \right\} = 0, \quad \forall\, i = \overline{1,n},\; n = \dim M.$$

Definition 99. The relativistic time-dependent semispray $\mathcal{S}_{\mathcal{E}\mathcal{D}} = (H, G)$ is called the *canonical semispray of the autonomous jet single-time Lagrange space of electrodynamics* $\mathcal{E}\mathcal{D}L_1^n$.

In the sequel, following our Riemann-Lagrange geometrical ideas from the preceding chapters, the canonical time-dependent semispray $\mathcal{S}_{\mathcal{E}\mathcal{D}} = (H, G)$ naturally induces on the 1-jet space $J^1(\mathbb{R}, M)$ a nonlinear connection

$$\Gamma_{\mathcal{E}\mathcal{D}} := \Gamma = \left(M_{(1)1}^{(i)}, N_{(1)j}^{(i)} \right), \tag{5.2}$$

which is called the *canonical nonlinear connection of the autonomous jet single-time Lagrange space of electrodynamics* $\mathcal{E}\mathcal{D}L_1^n$. In this way, we have the following:

Theorem 100. *The canonical nonlinear connection (5.2) of the autonomous jet single-time Lagrange space of electrodynamics* $\mathcal{E}\mathcal{D}L_1^n$ *is determined by the temporal components*

$$M_{(1)1}^{(i)} = 2H_{(1)1}^{(i)} = -\kappa_{11}^1 y_1^i \tag{5.3}$$

and the spatial components

$$N_{(1)j}^{(i)} = \frac{\partial G_{(1)1}^{(i)}}{\partial y_1^j} = \gamma_{jm}^i y_1^m + \frac{h_{11}\varphi^{im}}{4} U_{(m)j}^{(1)}. \tag{5.4}$$

Now, let us consider the adapted bases associated to the canonical nonlinear connection (5.2), which are given by

$$\left\{ \frac{\delta}{\delta t}, \frac{\delta}{\delta x^i}, \frac{\partial}{\partial y_1^i} \right\} \subset \mathcal{X}(J^1(\mathbb{R}, M))$$

and

$$\{dt, dx^i, \delta y_1^i\} \subset \mathcal{X}^*(J^1(\mathbb{R}, M)),$$

where

$$\frac{\delta}{\delta t} = \frac{\partial}{\partial t} - M_{(1)1}^{(m)} \frac{\partial}{\partial y_1^m},$$

$$\frac{\delta}{\delta x^i} = \frac{\partial}{\partial x^i} - N_{(1)i}^{(m)} \frac{\partial}{\partial y_1^m}, \tag{5.5}$$

$$\delta y_1^i = dy_1^i + M_{(1)1}^{(i)} dt + N_{(1)m}^{(i)} dx^m.$$

Following again the geometrical ideas from the preceding chapters, by direct computations, we can determine the adapted components of the *generalized Cartan canonical connection* of the jet autonomous single-time Lagrange space of electrodynamics $\mathcal{E}\mathcal{D}L_1^n$, together with its local d-torsions and d-curvatures.

Theorem 101. (i) *The generalized Cartan canonical connection*

$$C\Gamma = \left(H_{11}^1, G_{j1}^k, L_{jk}^i, C_{j(k)}^{i(1)} \right)$$

of the jet autonomous single-time Lagrange space of electrodynamics $\mathcal{EDL}_1^n$ has the following adapted components:

$$H_{11}^1 = \kappa_{11}^1, \quad G_{j1}^k = 0, \quad L_{jk}^i = \gamma_{jk}^i, \quad C_{j(k)}^{i(1)} = 0. \tag{5.6}$$

(ii) *The torsion* **T** *of the generalized Cartan canonical connection of the space $\mathcal{EDL}_1^n$ is determined by* **two** *effective adapted components, namely,*

$$R_{(1)1j}^{(k)} = -\frac{h_{11}\varphi^{km}}{4}\left[\kappa_{11}^1 U_{(m)j}^{(1)} + \frac{\partial U_{(m)j}^{(1)}}{\partial t}\right],$$

$$R_{(1)ij}^{(k)} = \mathfrak{R}_{mij}^k y_1^m + \frac{h_{11}\varphi^{km}}{4}\left[U_{(m)i|j}^{(1)} + U_{(m)j|i}^{(1)}\right], \tag{5.7}$$

where $\mathfrak{R}_{mij}^k(x)$ are the local curvature tensors of the semi-Riemannian metric $\varphi_{ij}(x)$ and "$_{|i}$" represents the local M-horizontal covariant derivative produced by the generalized Cartan canonical connection $C\Gamma$.

(iii) *The curvature* **R** *of the generalized Cartan canonical connection of the space $\mathcal{EDL}_1^n$ is determined by a* **single** *effective adapted component, namely,*

$$R_{(1)(i)jk}^{(l)(1)} = R_{ijk}^l = \mathfrak{R}_{ijk}^l,$$

that is exactly the curvature tensor of the semi-Riemannian metric $\varphi_{ij}(x)$.

5.2 GEOMETRICAL MAXWELL EQUATIONS ON $\mathcal{EDL}_1^N$

In order to describe our Riemann-Lagrange electromagnetic theory on the autonomous jet single-time Lagrange space of electrodynamics $\mathcal{EDL}_1^n$, let us remark that, by a simple direct calculation, we obtain the following:

Proposition 102. *The metrical deflection d-tensors of the space $\mathcal{EDL}_1^n$ are expressed by the formulas:*

$$\bar{D}_{(i)1}^{(1)} = \left[h^{11}\varphi_{im}y_1^m\right]_{/1} = 0,$$

$$D_{(i)j}^{(1)} = [h^{11}\varphi_{im}y_1^m]_{|j} = -\frac{1}{4}U_{(i)j}^{(1)}, \tag{5.1}$$

$$d_{(i)(j)}^{(1)(1)} = [h^{11}\varphi_{im}y_1^m]|_{(j)}^{(1)} = h^{11}\varphi_{ij},$$

where "$_{/1}$," "$_{|j}$," and "$|_{(j)}^{(1)}$" are the local covariant derivatives induced by the generalized Cartan canonical connection $C\Gamma$ of the space $\mathcal{EDL}_1^n$.

Moreover, taking into account the general formulas which give the electromagnetic components of a general relativistic time-dependent Lagrange space RRL_1^n, we find the following:

Theorem 103. *The local electromagnetic d-tensors of the autonomous single-time Lagrange space of electrodynamics $\mathcal{EDL}_1^n$ have the expressions*

$$
\begin{aligned}
F_{(i)j}^{(1)} &= \frac{1}{2}\left[D_{(i)j}^{(1)} - D_{(j)i}^{(1)}\right] = \frac{1}{8}\left[U_{(j)i}^{(1)} - U_{(i)j}^{(1)}\right] = -\frac{1}{4}U_{(i)j}^{(1)}, \\
f_{(i)(j)}^{(1)(1)} &= \frac{1}{2}\left[d_{(i)(j)}^{(1)(1)} - d_{(j)(i)}^{(1)(1)}\right] = 0.
\end{aligned}
\tag{5.2}
$$

Now, particularizing to $\mathcal{EDL}_1^n$ the geometrical Maxwell equations of the electromagnetic field of a general relativistic time-dependent Lagrange space RRL_1^n, we obtain the main result of the single-time electromagnetism on the space $\mathcal{EDL}_1^n$:

Theorem 104. *The electromagnetic components $F_{(i)j}^{(1)}$ of the autonomous single-time Lagrange space of electrodynamics $\mathcal{EDL}_1^n$ are governed by the following* **geometrical Maxwell equations***:*

$$
\begin{cases}
F_{(i)j/1}^{(1)} = \dfrac{1}{2}\mathcal{A}_{\{i,j\}}\left[h^{11}\varphi_{im}R_{(1)1j}^{(m)}\right] \\[2mm]
\sum_{\{i,j,k\}} F_{(i)j|k}^{(1)} = 0 \\[2mm]
\sum_{\{i,j,k\}} F_{(i)j}^{(1)}|_{(k)}^{(1)} = 0,
\end{cases}
\tag{5.3}
$$

where $\mathcal{A}_{\{i,j\}}$ represents an alternate sum and $\sum_{\{i,j,k\}}$ represents a cyclic sum.

5.3 GEOMETRICAL EINSTEIN EQUATIONS ON $\mathcal{EDL}_1^N$

To expose our generalized Riemann-Lagrange gravitational theory on the autonomous single-time Lagrange space of electrodynamics $\mathcal{EDL}_1^n$, we recall that the fundamental vertical metrical d-tensor of the single-time Lagrange space $\mathcal{EDL}_1^n$, given by

$$
G_{(i)(j)}^{(1)(1)} = h^{11}(t)\varphi_{ij}(x),
$$

and the canonical nonlinear connection (5.2), given by (5.3) and (5.4), produce a single-time gravitational h-potential $\mathbb{G}$ on the 1-jet space $J^1(\mathbb{R}, M)$, which is locally expressed by

$$
\mathbb{G} = h_{11}dt \otimes dt + \varphi_{ij}dx^i \otimes dx^j + h^{11}\varphi_{ij}\delta y_1^i \otimes \delta y_1^j.
\tag{5.1}
$$

In order to describe the local geometrical Einstein equations of the single-time gravitational h-potential $\mathbb{G}$ (together with their geometrical conservation laws) in the adapted basis

$$
\{X_A\} = \left\{\frac{\delta}{\delta t}, \frac{\delta}{\delta x^i}, \frac{\partial}{\partial y_1^i}\right\},
$$

let us consider

$$
C\Gamma = \left(\kappa_{11}^1, 0, \gamma_{jk}^i, 0\right),
$$

the generalized Cartan canonical connection of the space $\mathcal{E}\mathcal{D}L_1^n$.

Taking into account the expressions of the adapted curvature d-tensors of the space $\mathcal{E}\mathcal{D}L_1^n$, we immediately find the following:

Theorem 105. *The Ricci tensor* $\mathrm{Ric}(C\Gamma)$ *of the autonomous single-time Lagrange space of electrodynamics* $\mathcal{E}\mathcal{D}L_1^n$ *is characterized only by a* **single** *effective local Ricci d-tensor,*

$$R_{ij} := \mathfrak{R}_{ij} = \mathfrak{R}_{ijm}^m,$$

which is exactly the Ricci tensor of the semi-Riemannian metric $\varphi_{ij}(x)$.

Consequently, using the notation $\mathfrak{R} = \varphi^{ij}\mathfrak{R}_{ij}$, we deduce the following:

Theorem 106. *The scalar curvature* $\mathrm{Sc}(C\Gamma)$ *of the generalized Cartan connection* $C\Gamma$ *of the space* $\mathcal{E}\mathcal{D}L_1^n$ *has the expression*

$$\mathrm{Sc}(C\Gamma) := \mathfrak{R},$$

where $\mathfrak{R}$ *is the scalar curvature of the semi-Riemannian metric* $\varphi_{ij}(x)$.

Particularizing the geometrical Einstein equations and the geometrical conservation laws of a general relativistic time-dependent Lagrange space RRL_1^n, we can establish the main result of the geometrical gravitational theory on the jet autonomous single-time Lagrange space of electrodynamics $\mathcal{E}\mathcal{D}L_1^n$.

Theorem 107. (i) *The local* **geometrical Einstein equations**, *which govern the jet single-time gravitational h-potential* $\mathbb{G}$ *of the space* $\mathcal{E}\mathcal{D}L_1^n$, *have the form*

$$
\begin{cases}
-\dfrac{\mathfrak{R}}{2}h_{11} = \mathcal{K}\mathcal{T}_{11}, \\[2mm]
\mathfrak{R}_{ij} - \dfrac{\mathfrak{R}}{2}\varphi_{ij} = \mathcal{K}\mathcal{T}_{ij}, \\[2mm]
-\dfrac{\mathfrak{R}}{2}h^{11}\varphi_{ij} = \mathcal{K}\mathcal{T}_{(i)(j)}^{(1)(1)},
\end{cases}
\tag{5.2}
$$

$$
\begin{cases}
0 = \mathcal{T}_{1i}, \quad 0 = \mathcal{T}_{i1}, \quad 0 = \mathcal{T}_{(i)1}^{(1)}, \\[2mm]
0 = \mathcal{T}_{1(i)}^{(1)}, \quad 0 = \mathcal{T}_{i(j)}^{(1)}, \quad 0 = \mathcal{T}_{(i)j}^{(1)},
\end{cases}
\tag{5.3}
$$

where $\mathcal{T}_{AB}$, $A, B \in \left\{1, i, {}_{(i)}^{(1)}\right\}$, *are the adapted local components of the stress-energy d-tensor of matter* $\mathcal{T}$.

(ii) *The* **geometrical conservation laws** *of the geometrical Einstein equations of the space* $\mathcal{E}\mathcal{D}L_1^n$ *are expressed by the classical formulas*

$$\left[\mathfrak{R}_j^k - \frac{\mathfrak{R}}{2}\delta_j^k\right]_{|k} = 0,$$

where $\mathfrak{R}_j^k = \varphi^{km}\mathfrak{R}_{mj}$.

CHAPTER 6

JET LOCAL SINGLE-TIME FINSLER-LAGRANGE GEOMETRY FOR THE RHEONOMIC BERWALD-MOÓR METRIC OF ORDER THREE

It is a well known fact that, in order to create the Relativity Theory, Einstein was forced to use the Riemannian geometry instead of the classical Euclidean geometry, the first one representing the natural mathematical model for the local *isotropic* space-time. But, there are recent studies of physicists which suggest a *non-isotropic* perspective of the space-time. For example, according to D. G. Pavlov [80], the concept of inertial body mass infers the necessity of study of local non-isotropic spaces. Obviously, for the study of non-isotropic physical phenomena, the Finsler geometry is very useful as mathematical framework.

The studies of Russian scholars (Asanov [7], Mikhailov [54], and Garas'ko and Pavlov [36], [37]) emphasize the importance of the Finsler geometry which is characterized by the total equivalence in rights of all non-isotropic directions. For such a reason, the frameworks developed by Asanov and Pavlov underline the important role played by the Berwald-Moór metric models (whose classical Finsler geometry background was studied by Matsumoto and Shimada [53]), whose fundamental function has the form[2]

$$F : TM \to \mathbb{R}, \qquad F(y) = \left(y^1 y^2 ... y^n\right)^{\frac{1}{n}} \qquad (n \geq 2),$$

in the theory of space-time structure and gravitation, as well as in unified gauge field theories. Because any tangent direction can be related to the proper time of an inertial reference frame, Pavlov considers that it is appropriate to generically call such spaces as "multi-dimensional times" [79], [80].

For such geometrical and physical reasons, this Chapter is devoted to the development on the 1-jet space $J^1(\mathbb{R}, M^3)$ of the Finsler-Lagrange geometry (which serves as an extended model within a theoretical-geometric gravitational and electromagnetic field theory) for the *rheonomic Berwald-Moór metric of order three*,

$$\overset{\circ}{F} : J^1(\mathbb{R}, M^3) \to \mathbb{R}, \qquad \overset{\circ}{F}(t, y) = \sqrt{h^{11}(t)} \cdot \sqrt[3]{y_1^1 y_1^2 y_1^3},$$

where $h_{11}(t)$ is a Riemannian metric on $\mathbb{R}$ and $(t, x^1, x^2, x^3, y_1^1, y_1^2, y_1^3)$ are the coordinates of the 1-jet space $J^1(\mathbb{R}, M^3)$. Consequently, in the sequel, we apply the general geometrical results from Chapter 4 to the particular rheonomic Berwald-Moór metric of order three $\overset{\circ}{F}$, in order to obtain the so-called *jet local Finsler-Lagrange geometry of the three-dimensional time*.

6.1   PRELIMINARY NOTATIONS AND FORMULAS

Let $(\mathbb{R}, h_{11}(t))$ be a Riemannian manifold, where $\mathbb{R}$ is the set of real numbers. The Christoffel symbol of the Riemannian metric $h_{11}(t)$ is

$$\kappa_{11}^1 = \frac{h^{11}}{2} \frac{dh_{11}}{dt}, \qquad h^{11} = \frac{1}{h_{11}} > 0.$$

Let also M^3 be a real manifold of dimension three, whose local coordinates are (x^1, x^2, x^3). Let us consider the 1-jet space $J^1(\mathbb{R}, M^3)$, whose local coordinates are

$$(t, x^1, x^2, x^3, y_1^1, y_1^2, y_1^3).$$

These transform by the rules (the Einstein convention of summation is assumed)

$$\widetilde{t} = \widetilde{t}(t), \quad \widetilde{x}^p = \widetilde{x}^p(x^q), \quad \widetilde{y}_1^p = \frac{\partial \widetilde{x}^p}{\partial x^q} \frac{dt}{d\widetilde{t}} \cdot y_1^q, \quad p, q = \overline{1, 3}, \qquad (6.1)$$

where $d\widetilde{t}/dt \neq 0$ and rank $(\partial \widetilde{x}^p / \partial x^q) = 3$. We consider that the manifold M^3 is endowed with a tensor of kind $(0, 3)$, given by the local components $G_{pqr}(x)$, which is totally symmetric in the indices p, q and r. Suppose that the d-tensor

$$G_{ij1} = 6 G_{ijp} y_1^p,$$

---

[2]For n even, we either assume all the y-components as being positive, or, alternatively, consider the product below the root taken in absolute value.

is non-degenerate, that is, there exists the d-tensor G^{jk1} on $J^1(\mathbb{R}, M^3)$ such that $G_{ij1}G^{jk1} = \delta_i^k$.

In this geometrical context, if we use the notation $G_{111} = G_{pqr}y_1^p y_1^q y_1^r$, we can consider the *third-root Finsler-like function* [90], [12] (it is 1-positive homogenous in the variable y)

$$F(t, x, y) = \sqrt[3]{G_{pqr}(x)y_1^p y_1^q y_1^r} \cdot \sqrt{h^{11}(t)} = \sqrt[3]{G_{111}(x, y)} \cdot \sqrt{h^{11}(t)}, \qquad (6.2)$$

where the Finsler function F has as domain of definition all values (t, x, y) which verify the condition $G_{111}(x, y) \neq 0$. If we denote $G_{i11} = 3G_{ipq}y_1^p y_1^q$, then the 3-positive homogeneity of the "y-function" G_{111} (this is in fact a d-tensor on the 1-jet space $J^1(\mathbb{R}, M^3)$) leads to the equalities

$$G_{i11} = \frac{\partial G_{111}}{\partial y_1^i}, \quad G_{i11}y_1^i = 3G_{111}, \quad G_{ij1}y_1^j = 2G_{i11},$$

$$G_{ij1} = \frac{\partial G_{i11}}{\partial y_1^j} = \frac{\partial^2 G_{111}}{\partial y_1^i \partial y_1^j}, \quad G_{ij11}y_1^i y_1^j = 6G_{111}, \quad \frac{\partial G_{ij1}}{\partial y_1^k} = 6G_{ijk}.$$

The *fundamental metrical d-tensor* produced by F is given by the formula

$$g_{ij}(t, x, y) = \frac{h_{11}(t)}{2} \frac{\partial^2 F^2}{\partial y_1^i \partial y_1^j}.$$

By direct computations, the fundamental metrical d-tensor takes the form

$$g_{ij}(x, y) = \frac{G_{111}^{-1/3}}{3} \left[G_{ij1} - \frac{1}{3G_{111}} G_{i11}G_{j11} \right]. \qquad (6.3)$$

Moreover, taking into account that the d-tensor G_{ij1} is non-degenerate, we deduce that the matrix $g = (g_{ij})$ admits the inverse $g^{-1} = (g^{jk})$. The entries of the inverse matrix g^{-1} are

$$g^{jk} = 3G_{111}^{1/3} \left[G^{jk1} + \frac{G_1^j G_1^k}{3(G_{111} - \mathcal{G}_{111})} \right], \qquad (6.4)$$

where $G_1^j = G^{jp1}G_{p11}$ and $3\mathcal{G}_{111} = G^{pq1}G_{p11}G_{q11}$.

6.2   THE RHEONOMIC BERWALD-MOÓR METRIC OF ORDER THREE

Beginning with this Section we will focus only on *rheonomic Berwald-Moór metric of order three*, which is the Finsler-like metric (6.2) for the particular case

$$G_{pqr} = \begin{cases} \dfrac{1}{3!}, & \{p, q, r\} \text{ - distinct indices}, \\[2mm] 0, & \text{otherwise}. \end{cases}$$

Consequently, the rheonomic Berwald-Moór metric of order three is given by

$$\mathring{F}(t,y) = \sqrt{h^{11}(t)} \cdot \sqrt[3]{y_1^1 y_1^2 y_1^3}. \tag{6.5}$$

Moreover, using preceding notations and formulas, we obtain the following relations:

$$G_{111} = y_1^1 y_1^2 y_1^3, \quad G_{i11} = \frac{G_{111}}{y_1^i},$$

$$G_{ij1} = (1 - \delta_{ij}) \frac{G_{111}}{y_1^i y_1^j} \quad \text{(no sum by } i \text{ or } j\text{)},$$

where δ_{ij} is the Kronecker symbol. Because we have

$$\det (G_{ij1})_{i,j=\overline{1,3}} = 2G_{111} \neq 0,$$

we find

$$G^{jk1} = \frac{(1 - 2\delta^{jk})}{2G_{111}} y_1^j y_1^k \quad \text{(no sum by } j \text{ or } k\text{)}.$$

It follows that we have $\mathcal{G}_{111} = (1/2)G_{111}$ and $G_1^j = (1/2)y_1^j$.

Replacing now the preceding computed entities into formulas (6.3) and (6.4), we get

$$g_{ij} = \frac{(2 - 3\delta_{ij})}{9} \frac{G_{111}^{2/3}}{y_1^i y_1^j} \quad \text{(no sum by } i \text{ or } j\text{)} \tag{6.6}$$

and

$$g^{jk} = (2 - 3\delta^{jk})G_{111}^{-2/3} y_1^j y_1^k \quad \text{(no sum by } j \text{ or } k\text{)}. \tag{6.7}$$

Using the general formulas (4.7), (4.8), and (4.9), we find the following geometrical result:

Theorem 108. *For the rheonomic Berwald-Moór metric of order three (6.5), the* **energy action functional**

$$\mathring{\mathbb{E}}(t, x(t)) = \int_a^b \mathring{F}^2 \sqrt{h_{11}} dt = \int_a^b \sqrt[3]{\{y_1^1 y_1^2 y_1^3\}^2} \cdot h^{11} \sqrt{h_{11}} dt$$

produces on the 1-jet space $J^1(\mathbb{R}, M^3)$ *the* **canonical nonlinear connection**

$$\Gamma = \left(M_{(1)1}^{(i)} = -\kappa_{11}^1 y_1^i, \ N_{(1)j}^{(i)} = 0 \right). \tag{6.8}$$

Because the canonical nonlinear connection (6.8) has the spatial components equal to zero, it follows that our subsequent geometrical theory becomes trivial, in a way. For such a reason, in order to avoid the triviality of our theory and in order to have a certain kind of symmetry, we will use on the 1-jet space $J^1(\mathbb{R}, M^3)$, by an *a priori* definition, the following nonlinear connection (which does not curve the space):

$$\overset{\circ}{\Gamma} = \left(M^{(i)}_{(1)1} = -\kappa^1_{11} y^i_1, \ N^{(i)}_{(1)j} = -\frac{\kappa^1_{11}}{2} \delta^i_j \right). \tag{6.9}$$

Remark 109. The spatial components of the nonlinear connection (6.9), which are given on the local chart $\mathcal{U}$ by the functions

$$\overset{\circ}{N} = \left(N^{(i)}_{(1)j} = -\frac{\kappa^1_{11}}{2} \delta^i_j \right),$$

have not a global character on the 1-jet space $J^1(\mathbb{R}, M^3)$, but has only a local character. Consequently, taking into account the transformation rules (1.8) and (1.18), it follows that the spatial nonlinear connection $\overset{\circ}{N}$ has in the local chart $\widetilde{\mathcal{U}}$ the following components:

$$\widetilde{N}^{(k)}_{(1)l} = -\frac{\widetilde{\kappa}^1_{11}}{2} \delta^k_l + \frac{1}{2} \frac{d\widetilde{t}}{dt} \frac{d^2 t}{d\widetilde{t}^2} \delta^k_l + \frac{\partial \widetilde{x}^k}{\partial x^m} \frac{\partial^2 x^m}{\partial \widetilde{x}^l \partial \widetilde{x}^r} \widetilde{y}^r_1.$$

6.3 CARTAN CANONICAL LINEAR CONNECTION. d-TORSIONS AND d-CURVATURES

The importance of the nonlinear connection (6.9) is coming from the possibility of construction of the dual *adapted bases* of d-vector fields

$$\left\{ \frac{\delta}{\delta t} = \frac{\partial}{\partial t} + \kappa^1_{11} y^p_1 \frac{\partial}{\partial y^p_1} \ ; \ \frac{\delta}{\delta x^i} = \frac{\partial}{\partial x^i} + \frac{\kappa^1_{11}}{2} \frac{\partial}{\partial y^i_1} \ ; \ \frac{\partial}{\partial y^i_1} \right\} \subset \mathcal{X}(E) \tag{6.10}$$

and d-covector fields

$$\left\{ dt \ ; \ dx^i \ ; \ \delta y^i_1 = dy^i_1 - \kappa^1_{11} y^i_1 dt - \frac{\kappa^1_{11}}{2} dx^i \right\} \subset \mathcal{X}^*(E), \tag{6.11}$$

where $E = J^1(\mathbb{R}, M^3)$. Note that, under a change of coordinates (6.1), the elements of the adapted bases (6.10) and (6.11) transform as classical tensors. Consequently, all subsequent geometrical objects on the 1-jet space $J^1(\mathbb{R}, M^3)$ (such as Cartan canonical connection, torsion, curvature, etc.) will be described in local adapted components.

Using the general result of the Theorem 76, by direct computations, we can give the following important geometrical result:

Theorem 110. *The Cartan canonical $\overset{\circ}{\Gamma}$-linear connection, produced by the rheonomic Berwald-Moór metric of order three (6.5), has the following adapted local components:*

$$C\overset{\circ}{\Gamma} = \left(\kappa^1_{11}, \ G^k_{j1} = 0, \ L^i_{jk} = \frac{\kappa^1_{11}}{2} C^{i(1)}_{j(k)}, \ C^{i(1)}_{j(k)} \right),$$

where, if we use the notation

$$A^i_{jk} = \frac{3\delta^i_j + 3\delta^i_k + 3\delta_{jk} - 9\delta^i_j\delta_{jk} - 2}{9} \quad (no\ sum\ by\ i,\ j\ or\ k),$$

then

$$C^{i(1)}_{j(k)} = A^i_{jk} \cdot \frac{y^i_1}{y^j_1 y^k_1} \quad (no\ sum\ by\ i,\ j\ or\ k).$$

Proof: Via the Berwald-Moór derivative operators (6.10) and (6.11), we use the general formulas which give the adapted components of the Cartan canonical connection (see Theorem 76):

$$G^k_{j1} = \frac{g^{km}}{2}\frac{\delta g_{mj}}{\delta t}, \quad L^i_{jk} = \frac{g^{im}}{2}\left(\frac{\delta g_{jm}}{\delta x^k} + \frac{\delta g_{km}}{\delta x^j} - \frac{\delta g_{jk}}{\delta x^m}\right),$$

$$C^{i(1)}_{j(k)} = \frac{g^{im}}{2}\left(\frac{\partial g_{jm}}{\partial y^k_1} + \frac{\partial g_{km}}{\partial y^j_1} - \frac{\partial g_{jk}}{\partial y^m_1}\right) = \frac{g^{im}}{2}\frac{\partial g_{jk}}{\partial y^m_1}.$$

Remark 111. The below properties of the d-tensor $C^{i(1)}_{j(k)}$ are true (sum by m):

$$C^{i(1)}_{j(k)} = C^{i(1)}_{k(j)}, \quad C^{i(1)}_{j(m)}y^m_1 = 0, \quad C^{m(1)}_{j(m)} = 0. \tag{6.12}$$

For similar properties, see also the papers [12], [53], or [90].

Remark 112. The coefficients A^l_{ij} have the following values:

$$A^l_{ij} = \begin{cases} -\dfrac{2}{9}, & i \neq j \neq l \neq i, \\[2mm] \dfrac{1}{9}, & i = j \neq l \text{ or } i = l \neq j \text{ or } j = l \neq i, \\[2mm] -\dfrac{2}{9}, & i = j = l. \end{cases} \tag{6.13}$$

Theorem 113. *The Cartan canonical connection $C\overset{\circ}{\Gamma}$ of the rheonomic Berwald-Moór metric of order three (6.5) has* **three** *effective adapted local torsion d-tensors:*

$$P^{(k)\,(1)}_{(1)i(j)} = -\frac{\kappa^1_{11}}{2}C^{k(1)}_{i(j)}, \quad P^{k(1)}_{i(j)} = C^{k(1)}_{i(j)},$$

$$R^{(k)}_{(1)1j} = \frac{1}{2}\left[\frac{d\kappa^1_{11}}{dt} - \kappa^1_{11}\kappa^1_{11}\right]\delta^k_j.$$

Proof: A general h-normal Γ-linear connection on the 1-jet space $J^1(\mathbb{R}, M^3)$ is characterized by *eight* effective d-tensors of torsion (see Theorem 55). For our Cartan canonical connection $C\overset{\circ}{\Gamma}$ these reduce to the following *three* (the other five cancel):

$$P^{(k)\,(1)}_{(1)i(j)} = \frac{\partial N^{(k)}_{(1)i}}{\partial y^j_1} - L^k_{ji}, \quad R^{(k)}_{(1)1j} = \frac{\delta M^{(k)}_{(1)1}}{\delta x^j} - \frac{\delta N^{(k)}_{(1)j}}{\delta t}, \quad P^{k(1)}_{i(j)} = C^{k(1)}_{i(j)}.$$

Theorem 114. *The Cartan canonical connection $C\mathring{\Gamma}$ of the rheonomic Berwald-Moór metric of order three (6.5) has* **three** *effective adapted local curvature d-tensors:*

$$R^l_{ijk} = \frac{\kappa^1_{11}\kappa^1_{11}}{4} S^{l(1)(1)}_{i(j)(k)}, \quad P^{l\ (1)}_{ij(k)} = \frac{\kappa^1_{11}}{2} S^{l(1)(1)}_{i(j)(k)},$$

$$S^{l(1)(1)}_{i(j)(k)} = \frac{\partial C^{l(1)}_{i(j)}}{\partial y^k_1} - \frac{\partial C^{l(1)}_{i(k)}}{\partial y^j_1} + C^{m(1)}_{i(j)} C^{l(1)}_{m(k)} - C^{m(1)}_{i(k)} C^{l(1)}_{m(j)}.$$

Proof: A general h-normal Γ-linear connection on the 1-jet space $J^1(\mathbb{R}, M^3)$ is characterized by *five* effective d-tensors of curvature (see Theorem 58). For our Cartan canonical connection $C\mathring{\Gamma}$ these reduce to the following *three* (the other two cancel):

$$R^l_{ijk} = \frac{\delta L^l_{ij}}{\delta x^k} - \frac{\delta L^l_{ik}}{\delta x^j} + L^m_{ij} L^l_{mk} - L^m_{ik} L^l_{mj},$$

$$P^{l\ (1)}_{ij(k)} = \frac{\partial L^l_{ij}}{\partial y^k_1} - C^{l(1)}_{i(k)|j} + C^{l(1)}_{i(m)} P^{(m)\ (1)}_{(1)j(k)},$$

$$S^{l(1)(1)}_{i(j)(k)} = \frac{\partial C^{l(1)}_{i(j)}}{\partial y^k_1} - \frac{\partial C^{l(1)}_{i(k)}}{\partial y^j_1} + C^{m(1)}_{i(j)} C^{l(1)}_{m(k)} - C^{m(1)}_{i(k)} C^{l(1)}_{m(j)},$$

where

$$C^{l(1)}_{i(k)|j} = \frac{\delta C^{l(1)}_{i(k)}}{\delta x^j} + C^{m(1)}_{i(k)} L^l_{mj} - C^{l(1)}_{m(k)} L^m_{ij} - C^{l(1)}_{i(m)} L^m_{kj}.$$

Remark 115. The curvature d-tensor $S^{l(1)(1)}_{i(j)(k)}$ has the properties

$$S^{l(1)(1)}_{i(j)(k)} + S^{l(1)(1)}_{i(k)(j)} = 0, \quad S^{l(1)(1)}_{i(j)(j)} = 0 \text{ (no sum by } j\text{)}.$$

Theorem 116. *The expressions of the curvature d-tensor $S^{l(1)(1)}_{i(j)(k)}$ are as follows*

1. $S^{l(1)(1)}_{i(i)(k)} = -\dfrac{1}{9} \dfrac{y^l_1}{\left(y^i_1\right)^2 y^k_1}$ *($i \neq k \neq l \neq i$ and no sum by i);*

2. $S^{l(1)(1)}_{i(j)(i)} = \dfrac{1}{9} \dfrac{y^l_1}{\left(y^i_1\right)^2 y^j_1}$ *($i \neq j \neq l \neq i$ and no sum by i);*

3. $S^{i(1)(1)}_{i(j)(k)} = 0$ *($i \neq j \neq k \neq i$ and no sum by i);*

4. $S^{l(1)(1)}_{i(l)(k)} = \dfrac{1}{9 y^i_1 y^k_1}$ *($i \neq k \neq l \neq i$ and no sum by l);*

5. $S^{l(1)(1)}_{i(j)(l)} = -\dfrac{1}{9 y^i_1 y^j_1}$ *($i \neq j \neq l \neq i$ and no sum by l);*

6. $S_{i(i)(l)}^{l(1)(1)} = \dfrac{1}{9\left(y_1^i\right)^2}$ ($i \neq l$ and no sum by i or l);

7. $S_{i(l)(i)}^{l(1)(1)} = -\dfrac{1}{9\left(y_1^i\right)^2}$ ($i \neq l$ and no sum by i or l);

8. $S_{l(l)(k)}^{l(1)(1)} = 0$ ($k \neq l$ and no sum by l);

9. $S_{l(j)(l)}^{l(1)(1)} = 0$ ($j \neq l$ and no sum by l).

Proof: For $j \neq k$, the expression of the curvature tensor $S_{i(j)(k)}^{l(1)(1)}$ takes the form (no sum by i, j, k, or l, but with sum by m)

$$
S_{i(j)(k)}^{l(1)(1)} = \left[\frac{A_{ij}^l \delta_k^l}{y_1^i y_1^j} - \frac{A_{ik}^l \delta_j^l}{y_1^i y_1^k} \right] + \left[\frac{A_{ik}^l \delta_{ij} y_1^l}{\left(y_1^i\right)^2 y_1^k} - \frac{A_{ij}^l \delta_{ik} y_1^l}{\left(y_1^i\right)^2 y_1^j} \right]
$$
$$
+ \left[A_{ij}^m A_{mk}^l - A_{ik}^m A_{mj}^l \right] \frac{y_1^l}{y_1^i y_1^j y_1^k},
$$

where the coefficients A_{ij}^l are given by the relations (6.13).

6.4   GEOMETRICAL FIELD THEORIES PRODUCED BY THE RHEONOMIC BERWALD-MOÓR METRIC OF ORDER THREE

6.4.1   Geometrical gravitational theory

From a physical point of view, on the 1-jet space $J^1(\mathbb{R}, M^3)$, the rheonomic Berwald-Moór metric of order three (6.5) produces the adapted metrical d-tensor

$$
\mathbb{G} = h_{11} dt \otimes dt + g_{ij} dx^i \otimes dx^j + h^{11} g_{ij} \delta y_1^i \otimes \delta y_1^j, \tag{6.14}
$$

where g_{ij} is given by (6.6). This may be regarded as a "non-isotropic gravitational potential" [55]. In such a physical context, the nonlinear connection $\overset{\circ}{\Gamma}$ [given by (6.9) and used in the construction of the distinguished 1-forms δy_1^i] prescribes, probably, a kind of "interaction" between (t)-, (x)- and (y)-fields.

We postulate that the non-isotropic gravitational potential $\mathbb{G}$ is governed by the *geometrical Einstein equations*

$$
\mathrm{Ric}\left(C\overset{\circ}{\Gamma}\right) - \frac{\mathrm{Sc}\left(C\overset{\circ}{\Gamma}\right)}{2} \mathbb{G} = \mathcal{K}\mathcal{T}, \tag{6.15}
$$

where $\mathrm{Ric}\left(C\overset{\circ}{\Gamma}\right)$ is the *Ricci d-tensor* associated to the Cartan canonical connection $C\overset{\circ}{\Gamma}$, $\mathrm{Sc}\left(C\overset{\circ}{\Gamma}\right)$ is the *scalar curvature*, $\mathcal{K}$ is the *Einstein constant*, and $\mathcal{T}$ is the intrinsic *stress-energy d-tensor* of matter.

In this way, working with the adapted basis of vector fields (6.10), we can find the local geometrical Einstein equations for the rheonomic Berwald-Moór metric of order three (6.5). First, by direct computations, we find the following:

Proposition 117. *The Ricci d-tensor of the Cartan canonical connection $C\mathring{\Gamma}$ of the rheonomic Berwald-Moór metric of order three (6.5) has the following effective adapted local Ricci d-tensors:*

$$R_{ij} = R^m_{ijm} = \frac{\overset{1}{\kappa}_{11}\overset{1}{\kappa}_{11}}{4} S^{(1)(1)}_{(i)(j)}, \quad P^{(1)}_{i(j)} = P^{(1)}_{(i)j} = P^{m\,(1)}_{ij(m)} = \frac{\overset{1}{\kappa}_{11}}{2} S^{(1)(1)}_{(i)(j)},$$

$$S^{(1)(1)}_{(i)(j)} = S^{m(1)(1)}_{i(j)(m)} = \frac{3\delta_{ij} - 1}{9} \cdot \frac{1}{y^i_1 y^j_1} \ (no \ sum \ by \ i \ or \ j). \tag{6.16}$$

Remark 118. The local Ricci d-tensor $S^{(1)(1)}_{(i)(j)}$ has the following expression:

$$S^{(1)(1)}_{(i)(j)} = \begin{cases} -\dfrac{1}{9}\dfrac{1}{y^i_1 y^j_1}, & i \neq j, \\[2ex] \dfrac{2}{9}\dfrac{1}{\left(y^i_1\right)^2}, & i = j. \end{cases}$$

Remark 119. Using the last equality of (6.16) and the relation (6.7), we deduce that the following equality is true (sum by r):

$$S^{m11}_i \overset{def}{=} g^{mr} S^{(1)(1)}_{(r)(i)} = G^{-2/3}_{111} \cdot \frac{1 - 3\delta^m_i}{3} \cdot \frac{y^m_1}{y^i_1} \ (no \ sum \ by \ i \ or \ m). \tag{6.17}$$

Moreover, by a direct calculation, we obtain the equalities

$$\sum_{m,r=1}^3 S^{m11}_r C^{r(1)}_{i(m)} = 0, \quad \sum_{m=1}^3 \frac{\partial S^{m11}_i}{\partial y^m_1} = \frac{2}{3} \cdot \frac{1}{y^i_1} \cdot G^{-2/3}_{111}. \tag{6.18}$$

Proposition 120. *The scalar curvature of the Cartan canonical connection $C\mathring{\Gamma}$ of the rheonomic Berwald-Moór metric of order three (6.5) is given by*

$$Sc\left(C\mathring{\Gamma}\right) = -\frac{4h_{11} + \overset{1}{\kappa}_{11}\overset{1}{\kappa}_{11}}{2} \cdot G^{-2/3}_{111}.$$

Proof: The general formula for the scalar curvature of a Cartan connection is (see Proposition 90)

$$Sc\left(C\mathring{\Gamma}\right) = g^{pq} R_{pq} + h_{11} g^{pq} S^{(1)(1)}_{(p)(q)}.$$

Describing the global geometrical Einstein equations (6.15) in the adapted basis of vector fields (6.10), we find the following important geometrical and physical result:

Theorem 121. *The adapted local* **geometrical Einstein equations** *that govern the non-isotropic gravitational potential (6.14), produced by the rheonomic Berwald-Moór metric of order three (6.5) and the nonlinear connection (6.9), are given by*

$$
\begin{cases}
\xi_{11} \cdot G_{111}^{-2/3} \cdot h_{11} = T_{11}, \\[2mm]
\dfrac{\kappa_{11}^1 \kappa_{11}^1}{4\mathcal{K}} S_{(i)(j)}^{(1)(1)} + \xi_{11} \cdot G_{111}^{-2/3} \cdot g_{ij} = T_{ij}, \\[2mm]
\dfrac{1}{\mathcal{K}} S_{(i)(j)}^{(1)(1)} + \xi_{11} \cdot G_{111}^{-2/3} \cdot h^{11} \cdot g_{ij} = T_{(i)(j)}^{(1)(1)},
\end{cases}
\tag{6.19}
$$

$$
\begin{cases}
0 = T_{1i}, \qquad 0 = T_{i1}, \qquad\quad 0 = T_{(i)1}^{(1)}, \\[2mm]
0 = T_{1(i)}^{(1)}, \quad \dfrac{\kappa_{11}^1}{2\mathcal{K}} S_{(i)(j)}^{(1)(1)} = T_{i(j)}^{(1)}, \quad \dfrac{\kappa_{11}^1}{2\mathcal{K}} S_{(i)(j)}^{(1)(1)} = T_{(i)j}^{(1)},
\end{cases}
\tag{6.20}
$$

where

$$
\xi_{11} = \frac{4h_{11} + \kappa_{11}^1 \kappa_{11}^1}{4\mathcal{K}}.
\tag{6.21}
$$

Remark 122. The adapted local geometrical Einstein equations (6.19) and (6.20) impose as the stress-energy d-tensor of matter T to be symmetric. In other words, the stress-energy d-tensor of matter T must verify the local symmetry conditions

$$
T_{AB} = T_{BA}, \quad \forall\, A, B \in \left\{ 1,\ i,\ {}^{(1)}_{(i)} \right\}.
$$

By direct computations, the adapted local geometrical Einstein equations (6.19) and (6.20) imply the following identities of the stress-energy d-tensor (sum by r):

$$
T_1^1 \overset{def}{=} h^{11} T_{11} = \xi_{11} \cdot G_{111}^{-2/3}, \qquad T_1^m \overset{def}{=} g^{mr} T_{r1} = 0,
$$

$$
T_{(1)1}^{(m)} \overset{def}{=} h_{11} g^{mr} T_{(r)1}^{(1)} = 0, \qquad T_i^1 \overset{def}{=} h^{11} T_{1i} = 0,
$$

$$
T_i^m \overset{def}{=} g^{mr} T_{ri} = \frac{\kappa_{11}^1 \kappa_{11}^1}{4\mathcal{K}} S_i^{m11} + \xi_{11} \cdot G_{111}^{-2/3} \cdot \delta_i^m,
$$

$$
T_{(1)i}^{(m)} \overset{def}{=} h_{11} g^{mr} T_{(r)i}^{(1)} = \frac{h_{11} \kappa_{11}^1}{2\mathcal{K}} S_i^{m11}, \qquad T_{(i)}^{1(1)} \overset{def}{=} h^{11} T_{1(i)}^{(1)} = 0,
$$

$$
T_{(i)}^{m(1)} \overset{def}{=} g^{mr} T_{r(i)}^{(1)} = \frac{\kappa_{11}^1}{2\mathcal{K}} S_i^{m11},
$$

$$
T_{(1)(i)}^{(m)(1)} \overset{def}{=} h_{11} g^{mr} T_{(r)(i)}^{(1)(1)} = \frac{h_{11}}{\mathcal{K}} S_i^{m11} + \xi_{11} \cdot G_{111}^{-2/3} \cdot \delta_i^m,
$$

where the distinguished tensor S_i^{m11} is given by (6.17) and ξ_{11} is given by (6.21).

Theorem 123. *The stress-energy d-tensor of matter* T *must verify the following* **geometrical conservation laws** *(summation by* m*):*

$$
\left\{
\begin{array}{l}
T^1_{1/1} + T^m_{1|m} + T^{(m)}_{(1)1}\big|^{(1)}_{(m)} = \dfrac{\left(h^{11}\right)^2}{16\mathcal{K}} \dfrac{dh_{11}}{dt} \left[2\dfrac{d^2 h_{11}}{dt^2} - \dfrac{3}{h_{11}}\left(\dfrac{dh_{11}}{dt}\right)^2\right] \cdot G^{-2/3}_{111}, \\[3ex]
T^1_{i/1} + T^m_{i|m} + T^{(m)}_{(1)i}\big|^{(1)}_{(m)} = 0, \\[3ex]
T^{1(1)}_{(i)/1} + T^{m(1)}_{(i)|m} + T^{(m)(1)}_{(1)(i)}\big|^{(1)}_{(m)} = 0,
\end{array}
\right.
$$

(6.22)

where (summation by m *and* r*)*

$$
T^1_{1/1} \stackrel{def}{=} \frac{\delta T^1_1}{\delta t} + T^1_1 \kappa^1_{11} - T^1_1 \kappa^1_{11} = \frac{\delta T^1_1}{\delta t},
$$

$$
T^m_{1|m} \stackrel{def}{=} \frac{\delta T^m_1}{\delta x^m} + T^r_1 L^m_{rm} = \frac{\delta T^m_1}{\delta x^m},
$$

$$
T^{(m)}_{(1)1}\big|^{(1)}_{(m)} \stackrel{def}{=} \frac{\partial T^{(m)}_{(1)1}}{\partial y^m_1} + T^{(r)}_{(1)1} C^{m(1)}_{r(m)} = \frac{\partial T^{(m)}_{(1)1}}{\partial y^m_1},
$$

$$
T^1_{i/1} \stackrel{def}{=} \frac{\delta T^1_i}{\delta t} + T^1_i \kappa^1_{11} - T^1_r G^r_{i1} = \frac{\delta T^1_i}{\delta t} + T^1_i \kappa^1_{11},
$$

$$
T^m_{i|m} \stackrel{def}{=} \frac{\delta T^m_i}{\delta x^m} + T^r_i L^m_{rm} - T^m_r L^r_{im} = \frac{\kappa^1_{11}}{2} \frac{\partial T^m_i}{\partial y^m_1},
$$

$$
T^{(m)}_{(1)i}\big|^{(1)}_{(m)} \stackrel{def}{=} \frac{\partial T^{(m)}_{(1)i}}{\partial y^m_1} + T^{(r)}_{(1)i} C^{m(1)}_{r(m)} - T^{(m)}_{(1)r} C^{r(1)}_{i(m)} = \frac{\partial T^{(m)}_{(1)i}}{\partial y^m_1},
$$

$$
T^{1(1)}_{(i)/1} \stackrel{def}{=} \frac{\delta T^{1(1)}_{(i)}}{\delta t} + 2 T^{1(1)}_{(i)} \kappa^1_{11},
$$

$$
T^{m(1)}_{(i)|m} \stackrel{def}{=} \frac{\delta T^{m(1)}_{(i)}}{\delta x^m} + T^{r(1)}_{(i)} L^m_{rm} - T^{m(1)}_{(r)} L^r_{im} = \frac{\kappa^1_{11}}{2} \frac{\partial T^{m(1)}_{(i)}}{\partial y^m_1},
$$

$$
T^{(m)(1)}_{(1)(i)}\big|^{(1)}_{(m)} \stackrel{def}{=} \frac{\partial T^{(m)(1)}_{(1)(i)}}{\partial y^m_1} + T^{(r)(1)}_{(1)(i)} C^{m(1)}_{r(m)} - T^{(m)(1)}_{(1)(r)} C^{r(1)}_{i(m)} = \frac{\partial T^{(m)(1)}_{(1)(i)}}{\partial y^m_1}.
$$

Proof: The geometrical conservation laws (6.22) are provided by direct computations, using the relations (6.12) and (6.18).

6.4.2 Geometrical electromagnetic theory

Using the general relations (4.10), we find for the rheonomic Berwald-Moór metric of order three (6.5) and the nonlinear connection (6.9) the electromagnetic 2-form

$$\mathbb{F} := \overset{\circ}{\mathbb{F}} = 0.$$

In conclusion, our Berwald-Moór geometrical electromagnetic theory on the 1-jet space $J^1(\mathbb{R}, M^3)$ is trivial (this means that it contains tautological geometrical Maxwell equations). In our opinion, this fact suggests that the Berwald-Moór geometrical structure on the 1-jet space $J^1(\mathbb{R}, M^3)$ contains rather gravitational connotations than electromagnetic ones.

CHAPTER 7

JET LOCAL SINGLE-TIME FINSLER-LAGRANGE APPROACH FOR THE RHEONOMIC BERWALD-MOÓR METRIC OF ORDER FOUR

It is obvious that our natural physical intuition distinguishes four dimensions in a natural correspondence with the material reality. Consequently, the four-dimensionality plays a special role in almost all modern physical theories [54]. Thus, taking into account the non-isotropic perspective of the space-time endowed with the Berwald-Moór metric (together with its total equivalence of the non-isotropic directions involved), on the four-dimensional time (i.e., on the four-dimensional linear space endowed with the Berwald-Moór metric), Pavlov's framework [37], [80] offers specific physical-geometrical interpretations such as:

- physical events = points in the four-dimensional space;

- straight lines = shortest curves;

- intervals = distances between the points along of a straight line;

- light pyramids = light cones in a pseudo-Euclidian space.

In such a geometrical and physical context, the aim of this Chapter is to apply the general geometrical results from Chapter 4 to the *rheonomic Berwald-Moór metric*

I-st Edition. By Vladimir Balan and Mircea Neagu.
© 2011 John Wiley & Sons, Inc. Published 2011 John Wiley & Sons, Inc.

of order four [3]

$$\overset{\circ}{F} : J^1(\mathbb{R}, M^4) \to \mathbb{R}, \qquad \overset{\circ}{F}(t, y) = \sqrt{h^{11}(t)} \sqrt[4]{y_1^1 y_1^2 y_1^3 y_1^4},$$

where $h_{11}(t)$ is a Riemannian metric on $\mathbb{R}$ and $(t, x^1, x^2, x^3, x^4, y_1^1, y_1^2, y_1^3, y_1^4)$ are the coordinates of the 1-jet space $J^1(\mathbb{R}, M^4)$, in order to obtain the so-called *jet local Finsler-Lagrange geometry of the four-dimensional time*. As a consequence, a theoretic-geometric gravitational and electromagnetic field theory is produced from the given rheonomic Berwald-Moór metric of order four $\overset{\circ}{F}$. At the same time, at the end of this Chapter, a geometrical approach of the dynamics of plasma, regarded as an 1-jet medium endowed with the Berwald-Moór metric of order four $\overset{\circ}{F}$, is constructed.

7.1 PRELIMINARY NOTATIONS AND FORMULAS

Let $(\mathbb{R}, h_{11}(t))$ be a Riemannian manifold, where $\mathbb{R}$ is the set of real numbers. The Christoffel symbol of the Riemannian metric $h_{11}(t)$ is

$$\kappa_{11}^1 = \frac{h^{11}}{2} \frac{dh_{11}}{dt}, \qquad h^{11} = \frac{1}{h_{11}} > 0.$$

Let also M^4 be a real manifold of dimension four, whose local coordinates are (x^1, x^2, x^3, x^4). Let us consider the 1-jet space $J^1(\mathbb{R}, M^4)$, whose local coordinates are

$$(t, x^1, x^2, x^3, x^4, y_1^1, y_1^2, y_1^3, y_1^4).$$

These transform by the rules (the Einstein convention of summation is assumed)

$$\tilde{t} = \tilde{t}(t), \quad \tilde{x}^p = \tilde{x}^p(x^q), \quad \tilde{y}_1^p = \frac{\partial \tilde{x}^p}{\partial x^q} \frac{dt}{d\tilde{t}} \cdot y_1^q, \qquad p, q = \overline{1, 4}, \qquad (7.1)$$

where $d\tilde{t}/dt \neq 0$ and rank $(\partial \tilde{x}^p / \partial x^q) = 4$. We consider that the manifold M^4 is endowed with a tensor of kind $(0, 4)$, given by the local components $G_{pqrs}(x)$, which is totally symmetric in the indices p, q, r and s. Suppose that the d-tensor,

$$G_{ij11} = 12 G_{ijpq} y_1^p y_1^q,$$

is non-degenerate, that is, there exists the d-tensor G^{jk11} on $J^1(\mathbb{R}, M^4)$ such that $G_{ij11} G^{jk11} = \delta_i^k$.

In this geometrical context, if we use the notation $G_{1111} = G_{pqrs} y_1^p y_1^q y_1^r y_1^s$, we can consider the *fourth-root Finsler-like function* [90], [12] (it is 1-positive homogenous in the variable y):

$$F(t, x, y) = \sqrt[4]{G_{pqrs}(x) y_1^p y_1^q y_1^r y_1^s} \cdot \sqrt{h^{11}(t)} = \sqrt[4]{G_{1111}(x, y)} \cdot \sqrt{h^{11}(t)}, \quad (7.2)$$

---

[3]We assume all the y-components as being positive, or, alternatively, consider the product below the root taken in absolute value.

where the Finsler function F has as domain of definition all values (t, x, y) which verify the condition $G_{1111}(x, y) > 0$. If we denote $G_{i111} = 4G_{ipqr}(x)y_1^p y_1^q y_1^r$, then the 4-positive homogeneity of the "y-function" G_{1111} [this is in fact a d-tensor on $J^1(\mathbb{R}, M^4)$] leads to the equalities

$$G_{i111} = \frac{\partial G_{1111}}{\partial y_1^i}, \quad G_{i111}y_1^i = 4G_{1111}, \quad G_{ij11}y_1^j = 3G_{i111},$$

$$G_{ij11} = \frac{\partial G_{i111}}{\partial y_1^j} = \frac{\partial^2 G_{1111}}{\partial y_1^i \partial y_1^j}, \quad G_{ij11}y_1^i y_1^j = 12G_{1111}.$$

The *fundamental metrical d-tensor* produced by F is given by the formula

$$g_{ij}(t, x, y) = \frac{h_{11}(t)}{2} \frac{\partial^2 F^2}{\partial y_1^i \partial y_1^j}.$$

By direct computations, the fundamental metrical d-tensor takes the form

$$g_{ij}(x, y) = \frac{1}{4\sqrt{G_{1111}}} \left[G_{ij11} - \frac{1}{2G_{1111}} G_{i111}G_{j111} \right]. \tag{7.3}$$

Moreover, taking into account that the d-tensor G_{ij11} is non-degenerate, we deduce that the matrix $g = (g_{ij})$ admits the inverse $g^{-1} = (g^{jk})$. The entries of the inverse matrix g^{-1} are

$$g^{jk} = 4\sqrt{G_{1111}} \left[G^{jk11} + \frac{G_1^j G_1^k}{2(G_{1111} - \mathcal{G}_{1111})} \right], \tag{7.4}$$

where $G_1^j = G^{jp11}G_{p111}$ and $2\mathcal{G}_{1111} = G^{pq11}G_{p111}G_{q111}$.

7.2 THE RHEONOMIC BERWALD-MOÓR METRIC OF ORDER FOUR

Beginning with this Section we will focus only on the *rheonomic Berwald-Moór metric of order four*, which is the Finsler-like metric (7.2) for the particular case

$$G_{pqrs} = \begin{cases} \dfrac{1}{4!}, & \{p, q, r, s\} \text{ - distinct indices,} \\[2mm] 0, & \text{otherwise.} \end{cases}$$

Consequently, the rheonomic Berwald-Moór metric of order four is given by

$$\overset{\circ}{F}(t, y) = \sqrt{h^{11}(t)} \cdot \sqrt[4]{y_1^1 y_1^2 y_1^3 y_1^4}. \tag{7.5}$$

Moreover, using preceding notations and formulas, we obtain the following relations:

$$G_{1111} = y_1^1 y_1^2 y_1^3 y_1^4, \quad G_{i111} = \frac{G_{1111}}{y_1^i},$$

$$G_{ij11} = (1 - \delta_{ij}) \frac{G_{1111}}{y_1^i y_1^j} \text{ (no sum by } i \text{ or } j),$$

where δ_{ij} is the Kronecker symbol. Because we have

$$\det (G_{ij11})_{i,j=\overline{1,4}} = -3 (G_{1111})^2 \neq 0,$$

we find

$$G^{jk11} = \frac{(1 - 3\delta^{jk})}{3G_{1111}} y_1^j y_1^k \text{ (no sum by } j \text{ or } k).$$

It follows that we have $\mathcal{G}_{1111} = (2/3)G_{1111}$ and $G_1^j = (1/3)y_1^j$.

Replacing now the preceding computed entities into the formulas (7.3) and (7.4), we get

$$g_{ij} = \frac{(1 - 2\delta_{ij}) \sqrt{G_{1111}}}{8} \frac{1}{y_1^i y_1^j} \text{ (no sum by } i \text{ or } j) \tag{7.6}$$

and

$$g^{jk} = \frac{2(1 - 2\delta^{jk})}{\sqrt{G_{1111}}} y_1^j y_1^k \text{ (no sum by } j \text{ or } k). \tag{7.7}$$

Using the general formulas (4.7), (4.8), and (4.9), we find the following geometrical result:

Proposition 124. *For the rheonomic Berwald-Moór metric of order four (7.5), the* **energy action functional**

$$\mathbb{\mathring{E}}(t, x(t)) = \int_a^b \mathring{F}^2 \sqrt{h_{11}} dt = \int_a^b \sqrt{y_1^1 y_1^2 y_1^3 y_1^4} \cdot h^{11} \sqrt{h_{11}} dt$$

produces on the 1-jet space $J^1(\mathbb{R}, M^4)$ the **canonical nonlinear connection**

$$\Gamma = \left(M_{(1)1}^{(i)} = -\kappa_{11}^1 y_1^i, \ N_{(1)j}^{(i)} = 0 \right). \tag{7.8}$$

Because the canonical nonlinear connection (7.8) has the spatial components equal to zero, it follows that our subsequent geometrical theory becomes trivial, in a way. For such a reason, in order to avoid the triviality of our theory and in order to have a certain kind of symmetry, we will use on the 1-jet space $J^1(\mathbb{R}, M^4)$, by an *a priori* definition, the following nonlinear connection (which does not curve the space):

$$\mathring{\Gamma} = \left(M_{(1)1}^{(i)} = -\kappa_{11}^1 y_1^i, \ N_{(1)j}^{(i)} = -\frac{\kappa_{11}^1}{3} \delta_j^i \right). \tag{7.9}$$

Remark 125. The spatial components of the nonlinear connection (7.9), which are given on the local chart $\mathcal{U}$ by the functions

$$\mathring{N} = \left(N_{(1)j}^{(i)} = -\frac{\kappa_{11}^1}{3} \delta_j^i \right),$$

do not have a global character on the 1-jet space $J^1(\mathbb{R}, M^4)$, but have only a local character. Consequently, taking into account the transformation rules (1.8) and (1.18), it follows that the spatial nonlinear connection $\overset{\circ}{N}$ has in the local chart $\widetilde{\mathcal{U}}$ the following components:

$$\widetilde{N}^{(k)}_{(1)l} = -\frac{\widetilde{\kappa}^1_{11}}{3}\delta^k_l + \frac{1}{3}\frac{d\widetilde{t}}{dt}\frac{d^2t}{d\widetilde{t}^2}\delta^k_l + \frac{\partial\widetilde{x}^k}{\partial x^m}\frac{\partial^2 x^m}{\partial\widetilde{x}^l\partial\widetilde{x}^r}\widetilde{y}^r_1.$$

7.3 CARTAN CANONICAL LINEAR CONNECTION. d-TORSIONS AND d-CURVATURES

The importance of the nonlinear connection (7.9) is coming from the possibility of construction of the dual *adapted bases* of d-vector fields

$$\left\{\frac{\delta}{\delta t} = \frac{\partial}{\partial t} + \kappa^1_{11}y^p_1\frac{\partial}{\partial y^p_1} \; ; \; \frac{\delta}{\delta x^i} = \frac{\partial}{\partial x^i} + \frac{\kappa^1_{11}}{3}\frac{\partial}{\partial y^i_1} \; ; \; \frac{\partial}{\partial y^i_1}\right\} \subset \mathcal{X}(E) \quad (7.10)$$

and d-covector fields

$$\left\{dt \; ; \; dx^i \; ; \; \delta y^i_1 = dy^i_1 - \kappa^1_{11}y^i_1 dt - \frac{\kappa^1_{11}}{3}dx^i\right\} \subset \mathcal{X}^*(E), \quad (7.11)$$

where $E = J^1(\mathbb{R}, M^4)$. Note that, under a change of coordinates (7.1), the elements of the adapted bases (7.10) and (7.11) transform as classical tensors. Consequently, all subsequent geometrical objects on the 1-jet space $J^1(\mathbb{R}, M^4)$ (as Cartan canonical connection, torsion, curvature, etc.) will be described in local adapted components.

Using the general result of the Theorem 76, by direct computations, we can give the following important geometrical result:

Theorem 126. *The Cartan canonical $\overset{\circ}{\Gamma}$-linear connection, produced by the rheonomic Berwald-Moór metric of order four (7.5), has the following adapted local components:*

$$C\overset{\circ}{\Gamma} = \left(\kappa^1_{11}, \; G^k_{j1} = 0, \; L^i_{jk} = \frac{\kappa^1_{11}}{3}C^{i(1)}_{j(k)}, \; C^{i(1)}_{j(k)}\right), \quad (7.12)$$

where, if we use the notation

$$A^i_{jk} = \frac{2\delta^i_j + 2\delta^i_k + 2\delta_{jk} - 8\delta^i_j\delta_{jk} - 1}{8} \; \text{(no sum by } i, \; j \text{ or } k),$$

then

$$C^{i(1)}_{j(k)} = A^i_{jk} \cdot \frac{y^i_1}{y^j_1 y^k_1} \; \text{(no sum by } i, \; j \text{ or } k).$$

Proof: Via the Berwald-Moór derivative operators (7.10) and (7.11), we use the general formulas which give the adapted components of the Cartan canonical connection (see Theorem 76):

$$G^k_{j1} = \frac{g^{km}}{2}\frac{\delta g_{mj}}{\delta t}, \quad L^i_{jk} = \frac{g^{im}}{2}\left(\frac{\delta g_{jm}}{\delta x^k} + \frac{\delta g_{km}}{\delta x^j} - \frac{\delta g_{jk}}{\delta x^m}\right),$$

$$C^{i(1)}_{j(k)} = \frac{g^{im}}{2}\left(\frac{\partial g_{jm}}{\partial y^k_1} + \frac{\partial g_{km}}{\partial y^j_1} - \frac{\partial g_{jk}}{\partial y^m_1}\right) = \frac{g^{im}}{2}\frac{\partial g_{jm}}{\partial y^k_1}.$$

Remark 127. The below properties of the d-tensor $C^{i(1)}_{j(k)}$ are true (see also the papers [10] and [53]):

$$C^{i(1)}_{j(k)} = C^{i(1)}_{k(j)}, \quad C^{i(1)}_{j(m)}y^m_1 = 0, \quad C^{m(1)}_{j(m)} = 0 \text{ (sum by } m\text{)}. \tag{7.13}$$

Remark 128. The coefficients A^l_{ij} have the following values:

$$A^l_{ij} = \begin{cases} -\dfrac{1}{8}, & i \neq j \neq l \neq i, \\[2mm] \dfrac{1}{8}, & i = j \neq l \text{ or } i = l \neq j \text{ or } j = l \neq i, \\[2mm] -\dfrac{3}{8}, & i = j = l. \end{cases} \tag{7.14}$$

Theorem 129. *The Cartan canonical connection* $C\mathring{\Gamma}$ *of the rheonomic Berwald-Moór metric of order four (7.5) has* **three** *effective local torsion d-tensors:*

$$P^{(k)\,(1)}_{(1)i(j)} = -\frac{1}{3}\kappa^1_{11}C^{k(1)}_{i(j)}, \quad P^{k(1)}_{i(j)} = C^{k(1)}_{i(j)},$$

$$R^{(k)}_{(1)1j} = \frac{1}{3}\left[\frac{d\kappa^1_{11}}{dt} - \kappa^1_{11}\kappa^1_{11}\right]\delta^k_j.$$

Proof: A general h-normal Γ-linear connection on the 1-jet space $J^1(\mathbb{R}, M^4)$ is characterized by *eight* effective d-tensors of torsion (see Theorem 55). For our Cartan canonical connection $C\mathring{\Gamma}$ these reduce to the following *three* (the other five cancel):

$$P^{(k)\,(1)}_{(1)i(j)} = \frac{\partial N^{(k)}_{(1)i}}{\partial y^j_1} - L^k_{ji}, \quad R^{(k)}_{(1)1j} = \frac{\delta M^{(k)}_{(1)1}}{\delta x^j} - \frac{\delta N^{(k)}_{(1)j}}{\delta t}, \quad P^{k(1)}_{i(j)} = C^{k(1)}_{i(j)}.$$

Theorem 130. *The Cartan canonical connection* $C\mathring{\Gamma}$ *of the rheonomic Berwald-Moór metric of order four (7.5) has* **three** *effective local curvature d-tensors:*

$$R^l_{ijk} = \frac{1}{9}\kappa^1_{11}\kappa^1_{11}S^{l(1)(1)}_{i(j)(k)}, \quad P^{l\,(1)}_{ij(k)} = \frac{1}{3}\kappa^1_{11}S^{l(1)(1)}_{i(j)(k)},$$

$$S^{l(1)(1)}_{i(j)(k)} = \frac{\partial C^{l(1)}_{i(j)}}{\partial y^k_1} - \frac{\partial C^{l(1)}_{i(k)}}{\partial y^j_1} + C^{m(1)}_{i(j)}C^{l(1)}_{m(k)} - C^{m(1)}_{i(k)}C^{l(1)}_{m(j)}.$$

Proof: A general h-normal Γ-linear connection on the 1-jet space $J^1(\mathbb{R}, M^4)$ is characterized by *five* effective d-tensors of curvature (see Theorem 58). For our Cartan canonical connection $C\overset{\circ}{\Gamma}$ these reduce to the following *three* (the other two cancel):

$$R^l_{ijk} = \frac{\delta L^l_{ij}}{\delta x^k} - \frac{\delta L^l_{ik}}{\delta x^j} + L^m_{ij} L^l_{mk} - L^m_{ik} L^l_{mj},$$

$$P^{l\ (1)}_{ij(k)} = \frac{\partial L^l_{ij}}{\partial y^k_1} - C^{l(1)}_{i(k)|j} + C^{l(1)}_{i(m)} P^{(m)\ (1)}_{(1)j(k)},$$

$$S^{l(1)(1)}_{i(j)(k)} = \frac{\partial C^{l(1)}_{i(j)}}{\partial y^k_1} - \frac{\partial C^{l(1)}_{i(k)}}{\partial y^j_1} + C^{m(1)}_{i(j)} C^{l(1)}_{m(k)} - C^{m(1)}_{i(k)} C^{l(1)}_{m(j)},$$

where

$$C^{l(1)}_{i(k)|j} = \frac{\delta C^{l(1)}_{i(k)}}{\delta x^j} + C^{m(1)}_{i(k)} L^l_{mj} - C^{l(1)}_{m(k)} L^m_{ij} - C^{l(1)}_{i(m)} L^m_{kj}.$$

Remark 131. The curvature d-tensor $S^{l(1)(1)}_{i(j)(k)}$ has the properties

$$S^{l(1)(1)}_{i(j)(k)} + S^{l(1)(1)}_{i(k)(j)} = 0, \qquad S^{l(1)(1)}_{i(j)(j)} = 0 \text{ (no sum by } j\text{)}.$$

Theorem 132. *The following expressions of the curvature d-tensor $S^{l(1)(1)}_{i(j)(k)}$ hold true:*

1. $S^{l(1)(1)}_{i(j)(k)} = 0$ *for* $\{i, j, k, l\}$ *distinct indices;*

2. $S^{l(1)(1)}_{i(i)(k)} = -\dfrac{1}{16} \dfrac{y^l_1}{\left(y^i_1\right)^2 y^k_1}$ *($i \neq k \neq l \neq i$ and no sum by i);*

3. $S^{l(1)(1)}_{i(j)(i)} = \dfrac{1}{16} \dfrac{y^l_1}{\left(y^i_1\right)^2 y^j_1}$ *($i \neq j \neq l \neq i$ and no sum by i);*

4. $S^{i(1)(1)}_{i(j)(k)} = 0$ *($i \neq j \neq k \neq i$ and no sum by i);*

5. $S^{l(1)(1)}_{i(l)(k)} = \dfrac{1}{16 y^i_1 y^k_1}$ *($i \neq k \neq l \neq i$ and no sum by l);*

6. $S^{l(1)(1)}_{i(j)(l)} = -\dfrac{1}{16 y^i_1 y^j_1}$ *($i \neq j \neq l \neq i$ and no sum by l);*

7. $S^{l(1)(1)}_{i(i)(l)} = \dfrac{1}{8 \left(y^i_1\right)^2}$ *($i \neq l$ and no sum by i or l);*

8. $S^{l(1)(1)}_{i(l)(i)} = -\dfrac{1}{8 \left(y^i_1\right)^2}$ *($i \neq l$ and no sum by i or l);*

9. $S^{l(1)(1)}_{l(l)(k)} = 0$ *($k \neq l$ and no sum by l);*

10. $S^{l(1)(1)}_{l(j)(l)} = 0$ *($j \neq l$ and no sum by l).*

Proof: For $j \neq k$, the expression of the curvature tensor $S^{l(1)(1)}_{i(j)(k)}$ takes the form (no sum by i, j, k or l, but with sum by m)

$$
S^{l(1)(1)}_{i(j)(k)} = \left[\frac{A^l_{ij}\delta^l_k}{y^i_1 y^j_1} - \frac{A^l_{ik}\delta^l_j}{y^i_1 y^k_1} \right] + \left[\frac{A^l_{ik}\delta_{ij}y^l_1}{\left(y^i_1\right)^2 y^k_1} - \frac{A^l_{ij}\delta_{ik}y^l_1}{\left(y^i_1\right)^2 y^j_1} \right]
$$

$$
+ \left[A^m_{ij}A^l_{mk} - A^m_{ik}A^l_{mj} \right] \frac{y^l_1}{y^i_1 y^j_1 y^k_1},
$$

where the coefficients A^l_{ij} are given by the relations (7.14).

7.4 GEOMETRICAL GRAVITATIONAL THEORY PRODUCED BY THE RHEONOMIC BERWALD-MOÓR METRIC OF ORDER FOUR

From a physical point of view, on the 1-jet space $J^1(\mathbb{R}, M^4)$, the rheonomic Berwald-Moór metric of order four (7.5) produces the adapted metrical d-tensor

$$
\mathbb{G} = h_{11}dt \otimes dt + g_{ij}dx^i \otimes dx^j + h^{11}g_{ij}\delta y^i_1 \otimes \delta y^j_1, \tag{7.15}
$$

where g_{ij} is given by (7.6). This is regarded as a "non-isotropic gravitational potential," while the nonlinear connection $\overset{\circ}{\Gamma}$ [given by (7.9) and used in the construction of the distinguished 1-forms δy^i_1] is regarded as a kind of "interaction" between (t)-, (x)-, and (y)-fields.

We postulate that the non-isotropic gravitational potential $\mathbb{G}$ is governed by the *geometrical Einstein equations*

$$
\mathrm{Ric}\left(C\overset{\circ}{\Gamma} \right) - \frac{\mathrm{Sc}\left(C\overset{\circ}{\Gamma} \right)}{2}\mathbb{G} = \mathcal{K}\mathcal{T}, \tag{7.16}
$$

where $\mathrm{Ric}\left(C\overset{\circ}{\Gamma} \right)$ is the *Ricci d-tensor* associated to the Cartan canonical connection $C\overset{\circ}{\Gamma}$, $\mathrm{Sc}\left(C\overset{\circ}{\Gamma} \right)$ is the *scalar curvature*, $\mathcal{K}$ is the *Einstein constant*, and $\mathcal{T}$ is the intrinsic *stress-energy d-tensor* of matter.

In this way, working with the adapted basis of vector fields (7.10), we can find the local geometrical Einstein equations for the rheonomic Berwald-Moór metric of order four (7.5). Firstly, by direct computations, we find the following:

Proposition 133. *The Ricci d-tensor of the Cartan canonical connection $C\overset{\circ}{\Gamma}$ of the rheonomic Berwald-Moór metric of order four (7.5) has the following effective adapted local Ricci d-tensors:*

$$
R_{ij} = R^m_{ijm} = \frac{1}{9}\kappa^1_{11}\kappa^1_{11}S^{(1)(1)}_{(i)(j)},
$$

$$
P^{(1)}_{i(j)} = P^{(1)}_{(i)j} = P^{m\,(1)}_{ij(m)} = \frac{1}{3}\kappa^1_{11}S^{(1)(1)}_{(i)(j)}, \tag{7.17}
$$

$$
S^{(1)(1)}_{(i)(j)} = S^{m(1)(1)}_{i(j)(m)} = \frac{7\delta_{ij} - 1}{8}\frac{1}{y^i_1 y^j_1} \text{ (no sum by } i \text{ or } j\text{)}.
$$

Remark 134. The local Ricci d-tensor $S_{(i)(j)}^{(1)(1)}$ has the following expression:

$$
S_{(i)(j)}^{(1)(1)} = \begin{cases} -\dfrac{1}{8}\dfrac{1}{y_1^i y_1^j}, & i \neq j, \\[2mm] \dfrac{3}{4}\dfrac{1}{\left(y_1^i\right)^2}, & i = j. \end{cases}
$$

Remark 135. Using the third equality of (7.17) and the equality (7.7), we deduce that the following equality is true (sum by r):

$$
S_i^{m11} \overset{def}{=} g^{mr} S_{(r)(i)}^{(1)(1)} = \frac{5 - 14\delta_i^m}{4} \cdot \frac{1}{\sqrt{G_{1111}}} \cdot \frac{y_1^m}{y_1^i} \quad \text{(no sum by } i \text{ or } m). \tag{7.18}
$$

Moreover, by a direct calculation, we obtain the equalities

$$
\sum_{m,r=1}^{4} S_r^{m11} C_{i(m)}^{r(1)} = 0, \quad \sum_{m=1}^{4} \frac{\partial S_i^{m11}}{\partial y_1^m} = \frac{3}{\sqrt{G_{1111}}} \frac{1}{y_1^i}. \tag{7.19}
$$

Proposition 136. *The scalar curvature of the Cartan canonical connection $C\overset{\circ}{\Gamma}$ of the rheonomic Berwald-Moór metric of order four (7.5) is given by*

$$
Sc\left(C\overset{\circ}{\Gamma}\right) = -\frac{9h_{11} + \kappa_{11}^1 \kappa_{11}^1}{\sqrt{G_{1111}}}.
$$

Proof: The general formula for the scalar curvature of a Cartan connection is (see Proposition 90)

$$
Sc\left(C\overset{\circ}{\Gamma}\right) = g^{pq} R_{pq} + h_{11} g^{pq} S_{(p)(q)}^{(1)(1)}.
$$

Describing the global geometrical Einstein equations (7.16) in the adapted basis of vector fields (7.10), we find the following important geometrical and physical result:

Theorem 137. *The local adapted* **geometrical Einstein equations** *that govern the non-isotropic gravitational potential (7.15), produced by the rheonomic Berwald-Moór metric of order four (7.5) and the nonlinear connection (7.9), are given by*

$$
\begin{cases} \dfrac{\xi_{11} \cdot h_{11}}{\sqrt{G_{1111}}} = \mathcal{T}_{11}, \\[3mm] \dfrac{\kappa_{11}^1 \kappa_{11}^1}{9\mathcal{K}} S_{(i)(j)}^{(1)(1)} + \dfrac{\xi_{11}}{\sqrt{G_{1111}}} \cdot g_{ij} = \mathcal{T}_{ij}, \\[3mm] \dfrac{1}{\mathcal{K}} S_{(i)(j)}^{(1)(1)} + \dfrac{\xi_{11}}{\sqrt{G_{1111}}} \cdot h^{11} \cdot g_{ij} = \mathcal{T}_{(i)(j)}^{(1)(1)}, \end{cases} \tag{7.20}
$$

$$
\begin{cases} 0 = \mathcal{T}_{1i}, \quad 0 = \mathcal{T}_{i1}, \quad\quad\quad 0 = \mathcal{T}_{(i)1}^{(1)}, \\[3mm] 0 = \mathcal{T}_{1(i)}^{(1)}, \quad \dfrac{\kappa_{11}^1}{3\mathcal{K}} S_{(i)(j)}^{(1)(1)} = \mathcal{T}_{i(j)}^{(1)}, \quad \dfrac{\kappa_{11}^1}{3\mathcal{K}} S_{(i)(j)}^{(1)(1)} = \mathcal{T}_{(i)j}^{(1)}, \end{cases} \tag{7.21}
$$

where

$$\xi_{11} = \frac{9h_{11} + \kappa_{11}^1 \kappa_{11}^1}{2\mathcal{K}}. \tag{7.22}$$

Remark 138. The local geometrical Einstein equations (7.20) and (7.21) impose as the stress-energy d-tensor of matter $\mathcal{T}$ to be symmetric. In other words, the stress-energy d-tensor of matter $\mathcal{T}$ must verify the local symmetry conditions

$$\mathcal{T}_{AB} = \mathcal{T}_{BA}, \quad \forall A, B \in \left\{ 1, i, \genfrac{}{}{0pt}{}{(1)}{(i)} \right\}.$$

By direct computations, the adapted local geometrical Einstein equations (7.20) and (7.21) imply the following identities of the stress-energy d-tensor (sum by r):

$$\mathcal{T}_1^1 \stackrel{def}{=} h^{11}\mathcal{T}_{11} = \frac{\xi_{11}}{\sqrt{G_{1111}}}, \quad \mathcal{T}_1^m \stackrel{def}{=} g^{mr}\mathcal{T}_{r1} = 0,$$

$$\mathcal{T}_{(1)1}^{(m)} \stackrel{def}{=} h_{11}g^{mr}\mathcal{T}_{(r)1}^{(1)} = 0, \quad \mathcal{T}_i^1 \stackrel{def}{=} h^{11}\mathcal{T}_{1i} = 0,$$

$$\mathcal{T}_i^m \stackrel{def}{=} g^{mr}\mathcal{T}_{ri} = \frac{\kappa_{11}^1 \kappa_{11}^1}{9\mathcal{K}} S_i^{m11} + \frac{\xi_{11}}{\sqrt{G_{1111}}} \delta_i^m,$$

$$\mathcal{T}_{(1)i}^{(m)} \stackrel{def}{=} h_{11}g^{mr}\mathcal{T}_{(r)i}^{(1)} = \frac{h_{11}\kappa_{11}^1}{3\mathcal{K}} S_i^{m11}, \quad \mathcal{T}_{(i)}^{1(1)} \stackrel{def}{=} h^{11}\mathcal{T}_{1(i)}^{(1)} = 0,$$

$$\mathcal{T}_{(i)}^{m(1)} \stackrel{def}{=} g^{mr}\mathcal{T}_{r(i)}^{(1)} = \frac{\kappa_{11}^1}{3\mathcal{K}} S_i^{m11},$$

$$\mathcal{T}_{(1)(i)}^{(m)(1)} \stackrel{def}{=} h_{11}g^{mr}\mathcal{T}_{(r)(i)}^{(1)(1)} = \frac{h_{11}}{\mathcal{K}} S_i^{m11} + \frac{\xi_{11}}{\sqrt{G_{1111}}} \delta_i^m,$$

where the d-tensor S_i^{m11} is given by (7.18) and ξ_{11} is given by (7.22).

Corollary 139. *The stress-energy d-tensor of matter $\mathcal{T}$ must verify the following* **geometrical conservation laws** *(summation by m):*

$$\left\{ \begin{array}{l} \mathcal{T}_{1/1}^1 + \mathcal{T}_{1|m}^m + \mathcal{T}_{(1)1}^{(m)}|_{(m)}^{(1)} = \dfrac{(h^{11})^2}{8\mathcal{K}} \dfrac{dh_{11}}{dt} \left[2\dfrac{d^2h_{11}}{dt^2} - \dfrac{3}{h_{11}} \left(\dfrac{dh_{11}}{dt} \right)^2 \right] \cdot \dfrac{1}{\sqrt{G_{1111}}}, \\[3ex] \mathcal{T}_{i/1}^1 + \mathcal{T}_{i|m}^m + \mathcal{T}_{(1)i}^{(m)}|_{(m)}^{(1)} = \dfrac{\kappa_{11}^1 \xi_{11}}{18} \cdot \dfrac{1}{\sqrt{G_{1111}}} \cdot \dfrac{1}{y_1^i}, \\[3ex] \mathcal{T}_{(i)/1}^{1(1)} + \mathcal{T}_{(i)|m}^{m(1)} + \mathcal{T}_{(1)(i)}^{(m)(1)}|_{(m)}^{(1)} = \dfrac{\xi_{11}}{6} \cdot \dfrac{1}{\sqrt{G_{1111}}} \cdot \dfrac{1}{y_1^i}, \end{array} \right.$$
$$\tag{7.23}$$

where (summation by m and r)

$$T^1_{1/1} \overset{def}{=} \frac{\delta T^1_1}{\delta t} + T^1_1 \kappa^1_{11} - T^1_1 \kappa^1_{11} = \frac{\delta T^1_1}{\delta t},$$

$$T^m_{1|m} \overset{def}{=} \frac{\delta T^m_1}{\delta x^m} + T^r_1 L^m_{rm} = \frac{\delta T^m_1}{\delta x^m},$$

$$T^{(m)}_{(1)1}\big|^{(1)}_{(m)} \overset{def}{=} \frac{\partial T^{(m)}_{(1)1}}{\partial y^m_1} + T^{(r)}_{(1)1} C^{m(1)}_{r(m)} = \frac{\partial T^{(m)}_{(1)1}}{\partial y^m_1},$$

$$T^1_{i/1} \overset{def}{=} \frac{\delta T^1_i}{\delta t} + T^1_i \kappa^1_{11} - T^1_r G^r_{i1} = \frac{\delta T^1_i}{\delta t} + T^1_i \kappa^1_{11},$$

$$T^m_{i|m} \overset{def}{=} \frac{\delta T^m_i}{\delta x^m} + T^r_i L^m_{rm} - T^m_r L^r_{im} = \frac{\kappa^1_{11}}{3} \frac{\partial T^m_i}{\partial y^m_1},$$

$$T^{(m)}_{(1)i}\big|^{(1)}_{(m)} \overset{def}{=} \frac{\partial T^{(m)}_{(1)i}}{\partial y^m_1} + T^{(r)}_{(1)i} C^{m(1)}_{r(m)} - T^{(m)}_{(1)r} C^{r(1)}_{i(m)} = \frac{\partial T^{(m)}_{(1)i}}{\partial y^m_1},$$

$$T^{1(1)}_{(i)/1} \overset{def}{=} \frac{\delta T^{1(1)}_{(i)}}{\delta t} + 2 T^{1(1)}_{(i)} \kappa^1_{11},$$

$$T^{m(1)}_{(i)|m} \overset{def}{=} \frac{\delta T^{m(1)}_{(i)}}{\delta x^m} + T^{r(1)}_{(i)} L^m_{rm} - T^{m(1)}_{(r)} L^r_{im} = \frac{\kappa^1_{11}}{3} \frac{\partial T^{m(1)}_{(i)}}{\partial y^m_1},$$

$$T^{(m)(1)}_{(1)(i)}\big|^{(1)}_{(m)} \overset{def}{=} \frac{\partial T^{(m)(1)}_{(1)(i)}}{\partial y^m_1} + T^{(r)(1)}_{(1)(i)} C^{m(1)}_{r(m)} - T^{(m)(1)}_{(1)(r)} C^{r(1)}_{i(m)} = \frac{\partial T^{(m)(1)}_{(1)(i)}}{\partial y^m_1}.$$

Proof: The geometrical conservation laws (7.23) are provided by direct computations, using the relations (7.13) and (7.19).

7.5 SOME PHYSICAL REMARKS AND COMMENTS

7.5.1 On gravitational theory

It is known that in the classical Relativity Theory of Einstein (which characterizes the gravity in an isotropic space-time) the tensor of matter must verify the conservation laws

$$T^m_{i;m} = 0, \quad \forall\, i = \overline{1,4},$$

where ";" means the covariant derivative produced by the Levi-Civita connection associated to semi-Riemannian metric $g_{ij}(x)$ (the gravitational potentials).

Comparatively, in our non-isotropic gravitational theory [with respect to the rheonomic Berwald-Moór metric of order four (7.5)] the conservation laws are replaced with ($i = \overline{1,4}$)

$$T_1 = \frac{(h^{11})^2}{8\mathcal{K}} \frac{dh_{11}}{dt} \left[2\frac{d^2 h_{11}}{dt^2} - \frac{3}{h_{11}} \left(\frac{dh_{11}}{dt} \right)^2 \right] \cdot \frac{1}{\sqrt{G_{1111}}},$$

$$T_i = \frac{\kappa^1_{11}\xi_{11}}{18} \cdot \frac{1}{\sqrt{G_{1111}}} \cdot \frac{1}{y^i_1}, \quad T^{(1)}_{(i)} = \frac{\xi_{11}}{6} \cdot \frac{1}{\sqrt{G_{1111}}} \cdot \frac{1}{y^i_1},$$

where

$$T_1 \stackrel{def}{=} T^1_{1/1} + T^m_{1|m} + T^{(m)}_{(1)1}|^{(1)}_{(m)},$$

$$T_i \stackrel{def}{=} T^1_{i/1} + T^m_{i|m} + T^{(m)}_{(1)i}|^{(1)}_{(m)},$$

$$T^{(1)}_{(i)} \stackrel{def}{=} T^{1(1)}_{(i)/1} + T^{m(1)}_{(i)|m} + T^{(m)(1)}_{(1)(i)}|^{(1)}_{(m)}.$$

By analogy with Einstein's theory, if we impose the conditions ($\forall\, i = \overline{1,4}$)

$$\begin{cases} T_1 = 0, \\ T_i = 0, \\ T^{(1)}_{(i)} = 0, \end{cases}$$

then we reach to the system of differential equations

$$\begin{cases} \dfrac{dh_{11}}{dt} \left[2\dfrac{d^2 h_{11}}{dt^2} - \dfrac{3}{h_{11}} \left(\dfrac{dh_{11}}{dt} \right)^2 \right] = 0, \\ 9h_{11} + \kappa^1_{11}\kappa^1_{11} = 0. \end{cases} \qquad (7.24)$$

Obviously, because we have $h_{11} > 0$, we deduce that the ODE system (7.24) does not have any solution. Consequently, we always have

$$[T_1]^2 + [T_i]^2 + \left[T^{(1)}_{(i)} \right]^2 \neq 0, \quad \forall\, i = \overline{1,4}.$$

In our opinion, this fact suggests that our geometrical gravitational theory [produced by the rheonomic Berwald-Moór gravitational potential (7.15)] is not suitable for media whose stress-energy d-components are

$$T_{AB} = 0, \quad \forall\, A, B \in \left\{ 1,\, i,\, \begin{smallmatrix}(1)\\(i)\end{smallmatrix} \right\}.$$

However, it is important to note that at "infinity"

(this means that $y^i_1 \to \infty, \quad \forall\, i = \overline{1,4}$),

our Berwald-Moór geometrical gravitational theory seems to be appropriate even for media characterized by a null stress-energy d-tensor of matter. This is because at "infinity" the stress-energy local d-tensors tend to become zero.

7.5.2   On electromagnetic theory

Using the general formula (4.10) in our particular case of the rheonomic Berwald-Moór metric of order four (7.5) and the nonlinear connection (7.9), we have the electromagnetic 2-form $\mathbb{F} := \overset{\circ}{\mathbb{F}} = 0$. Consequently, our Berwald-Moór geometrical electromagnetic theory of order four is again trivial (this means that we have tautological geometrical Maxwell equations). In our opinion, this fact also suggests that the rheonomic Berwald-Moór metric of order four (7.5) has rather strong gravitational connotations than electromagnetic ones.

7.6   GEOMETRIC DYNAMICS OF PLASMA IN JET SPACES WITH RHEONOMIC BERWALD-MOÓR METRIC OF ORDER FOUR

7.6.1   Introduction

During the so-called "radiation epoch" (in which photons are strongly coupled with the matter), the interactions between the various constituents of the Universal matter include radiation-plasma coupling, which is described by the plasma dynamics. Although it is not traditional to characterize the radiation epoch by the dominance of plasma interactions, however, it may be also called the plasma epoch (see [43]). This is because, in the plasma epoch, the electromagnetic interaction dominates all four fundamental physical forces (electrical, magnetic, gravitational, and nuclear).

Nowadays, the Plasma Physics is a well established field of Theoretical Physics, although the formulation of magnetohydrodynamics (MHD) in a curved space-time is a relatively new development (see Punsly [86]). The MHD processes in an isotropic space-time are intensively studied by many physicists. For example, the MHD equations in an expanding Universe are investigated by Kleidis, Kuiroukidis, D. Papadopoulos, and Vlahos in [43]. Considering the interaction of the gravitational waves with the plasma in the presence of a weak magnetic field, D. B. Papadopoulos also investigates the relativistic hydromagnetic equations [78]. The electromagnetic-gravitational dynamics into plasmas with pressure and viscosity is studied by Das, DeBenedictis, Kloster, and Tariq in the paper [32].

It is important to note that all preceding physical studies are done on an isotropic four-dimensional space-time, represented by a semi- (pseudo-) Riemannian space with the signature $(+, +, +, -)$. Consequently, the Riemannian geometrical methods are used as a pattern over there.

From a geometrical point of view, using the Finslerian geometrical methods, the plasma dynamics was extended on non-isotropic space-times by V. Gîrţu and Ciubotariu in the paper [40]. More generally, after the development of Lagrangian geometry of the tangent bundle by Miron and Anastasiei [55], the generalized Lagrange geometrical objects describing the relativistic magnetized plasma were studied by M. Gîrţu, V. Gîrţu, and Postolache in the paper [41].

In such a physical perspective and because of all preceding geometrical and physical reasons, this paper is devoted to the development on the 1-jet space $J^1(\mathbb{R}, M^4)$ of the geometric dynamics of plasma endowed with the *relativistic rheonomic Berwald-*

Moór metric of order four

$$\overset{\circ}{F} : J^1(\mathbb{R}, M^4) \to \mathbb{R}, \qquad \overset{\circ}{F}(t, y) = \sqrt{h^{11}(t)} \sqrt[4]{y_1^1 y_1^2 y_1^3 y_1^4},$$

where $h_{11}(t)$ is a Riemannian metric on $\mathbb{R}$ and $(t, x^1, x^2, x^3, x^4, y_1^1, y_1^2, y_1^3, y_1^4)$ are the coordinates on the 1-jet space $J^1(\mathbb{R}, M^4)$.

7.6.2 Generalized Lagrange geometrical approach of the non-isotropic plasma on 1-jet spaces

Let $(\mathbb{R}, h_{11}(t))$ be the set of real numbers endowed with a Riemannian structure. Let us suppose that the Christoffel symbol of the Riemannian metric $h_{11}(t)$ is

$$\kappa_{11}^1 = \frac{h^{11}}{2} \frac{dh_{11}}{dt}, \qquad h^{11} = \frac{1}{h_{11}} > 0.$$

Let us consider that M^n is a real manifold of dimension n, whose local coordinates are $\left(x^i\right)_{i=\overline{1,n}}$. Note that, in this Subsection, the Latin letters run from 1 to n, and the Einstein convention of summation is assumed. Let $J^1(\mathbb{R}, M^n)$ be the 1-jet space of dimension $2n + 1$, whose local coordinates are (t, x^i, y_1^i). These transform by the rules

$$\begin{cases} \widetilde{t} = \widetilde{t}(t), \\[2mm] \widetilde{x}^i = \widetilde{x}^i(x^j), \\[2mm] \widetilde{y}_1^i = \dfrac{\partial \widetilde{x}^i}{\partial x^j} \dfrac{dt}{d\widetilde{t}} y_1^j, \end{cases}$$

where $d\widetilde{t}/dt \neq 0$ and rank $(\partial \widetilde{x}^p / \partial x^q) = n$.

Let $RRGL_1^n = \left(J^1(\mathbb{R}, M^n), G_{(i)(j)}^{(1)(1)} = h^{11} g_{ij}\right)$ be a *generalized relativistic time-dependent Lagrange space* (for more details, see Neagu [65]), where $g_{ij}(t, x^k, y_1^k)$ is a metrical d-tensor on $J^1(\mathbb{R}, M^n)$ (not necessarily provided by a jet Lagrangian function), which is symmetrical, non-degenerate, and has a constant signature.

Let us consider that $RRGL_1^n$ is endowed with a nonlinear connection having the form

$$\Gamma = \left(M_{(1)1}^{(i)} = -\kappa_{11}^1 y_1^i, \; N_{(1)j}^{(i)}\right).$$

The nonlinear connection Γ produces on $J^1(\mathbb{R}, M^n)$ the following dual adapted bases of d-vectors and d-covectors:

$$\left\{\frac{\delta}{\delta t}, \frac{\delta}{\delta x^i}, \frac{\partial}{\partial y_1^i}\right\} \subset \mathcal{X}(J^1(\mathbb{R}, M^n)), \qquad \left\{dt, dx^i, \delta y_1^i\right\} \subset \mathcal{X}^*(J^1(\mathbb{R}, M^n)),$$

where

$$\frac{\delta}{\delta t} = \frac{\partial}{\partial t} + \kappa_{11}^1 y_1^m \frac{\partial}{\partial y_1^m}, \qquad \frac{\delta}{\delta x^i} = \frac{\partial}{\partial x^i} - N_{(1)i}^{(m)} \frac{\partial}{\partial y_1^m},$$

$$\delta y_1^i = dy_1^i - \kappa_{11}^1 y_1^i dt + N_{(1)m}^{(i)} dx^m.$$

Moreover, it is obvious that the generalized relativistic time-dependent Lagrange space $RRGL_1^n$ produces on the 1-jet space $J^1(\mathbb{R}, M^n)$ the global metrical d-tensor

$$\mathbb{G} = h_{11}dt \otimes dt + g_{ij}dx^i \otimes dx^j + h^{11}g_{ij}\delta y_1^i \otimes \delta y_1^j,$$

which is endowed with the physical meaning of non-isotropic gravitational potential. Obviously, the d-tensor $\mathbb{G}$ has the adapted components

$$\mathbb{G}_{AB} = \begin{cases} h_{11}, & \text{for} & A = 1, \quad B = 1, \\ g_{ij}, & \text{for} & A = i, \quad B = j, \\ h^{11}g_{ij}, & \text{for} & A = \overset{(1)}{(i)}, \quad B = \overset{(1)}{(j)}, \\ 0, & \text{otherwise.} \end{cases}$$

Following the geometrical ideas from the previous chapters, the above geometrical ingredients lead us to the Cartan canonical Γ-linear connection (see Theorem 76)

$$C\Gamma = \left(\kappa_{11}^1, \; G_{j1}^k, \; L_{jk}^i, \; C_{j(k)}^{i(1)} \right),$$

where

$$G_{j1}^k = \frac{g^{km}}{2}\frac{\delta g_{mj}}{\delta t}, \qquad L_{jk}^i = \frac{g^{im}}{2}\left(\frac{\delta g_{jm}}{\delta x^k} + \frac{\delta g_{km}}{\delta x^j} - \frac{\delta g_{jk}}{\delta x^m} \right),$$

$$C_{j(k)}^{i(1)} = \frac{g^{im}}{2}\left(\frac{\partial g_{jm}}{\partial y_1^k} + \frac{\partial g_{km}}{\partial y_1^j} - \frac{\partial g_{jk}}{\partial y_1^m} \right). \tag{7.25}$$

In the sequel, the Cartan linear connection $C\Gamma$, given by (7.25), induces the $\mathbb{R}$-horizontal ($h_{\mathbb{R}}-$) covariant derivative

$$\begin{aligned} D_{1k(1)(l)\dots/1}^{1i(j)(1)\dots} =& \frac{\delta D_{1k(1)(l)\dots}^{1i(j)(1)\dots}}{\delta t} + D_{1k(1)(l)\dots}^{1i(j)(1)\dots}\kappa_{11}^1 + D_{1k(1)(l)\dots}^{1m(j)(1)\dots}G_{m1}^i \\ &+ D_{1k(1)(l)\dots}^{1i(m)(1)\dots}G_{m1}^j + D_{1k(1)(l)\dots}^{1i(j)(1)\dots}\kappa_{11}^1 + \dots \\ &- D_{1k(1)(l)\dots}^{1i(j)(1)\dots}\kappa_{11}^1 - D_{1m(1)(l)\dots}^{1i(j)(1)\dots}G_{k1}^m \\ &- D_{1k(1)(l)\dots}^{1i(j)(1)\dots}\kappa_{11}^1 - D_{1k(1)(m)\dots}^{1i(j)(1)\dots}G_{l1}^m \dots, \end{aligned}$$

the M-horizontal (h_M-) covariant derivative

$$\begin{aligned} D_{1k(1)(l)\dots|p}^{1i(j)(1)\dots} =& \frac{\delta D_{1k(1)(l)\dots}^{1i(j)(1)\dots}}{\delta x^p} + D_{1k(1)(l)\dots}^{1m(j)(1)\dots}L_{mp}^i + D_{1k(1)(l)\dots}^{1i(m)(1)\dots}L_{mp}^j + \dots \\ &- D_{1m(1)(l)\dots}^{1i(j)(1)\dots}L_{kp}^m - D_{1k(1)(m)\dots}^{1i(j)(1)\dots}L_{lp}^m - \dots, \end{aligned}$$

and the vertical ($v-$) covariant derivative

$$\begin{aligned} D_{1k(1)(l)\dots|(p)}^{1i(j)(1)\dots(1)} =& \frac{\partial D_{1k(1)(l)\dots}^{1i(j)(1)\dots}}{\partial y_1^p} + D_{1k(1)(l)\dots}^{1m(j)(1)\dots}C_{m(p)}^{i(1)} + D_{1k(1)(l)\dots}^{1i(m)(1)\dots}C_{m(p)}^{j(1)} + \dots \\ &- D_{1m(1)(l)\dots}^{1i(j)(1)\dots}C_{k(p)}^{m(1)} - D_{1k(1)(m)\dots}^{1i(j)(1)\dots}C_{l(p)}^{m(1)} - \dots, \end{aligned}$$

where

$$D = D_{1k(1)(l)...}^{1i(j)(1)...}(t, x^r, y_1^r) \frac{\delta}{\delta t} \otimes \frac{\delta}{\delta x^i} \otimes \frac{\partial}{\partial y_1^j} \otimes dt \otimes dx^k \otimes \delta y_1^l \otimes ...$$

is an arbitrary d-tensor on $J^1(\mathbb{R}, M)$.

Remark 140. We recall that the Cartan covariant derivatives produced by $C\Gamma$ have the metrical properties

$$h_{11/1} = h^{11}_{/1} = 0, \quad h_{11|k} = h^{11}_{|k} = 0, \quad h_{11}|_{(k)}^{(1)} = h^{11}|_{(k)}^{(1)} = 0,$$

$$g_{ij/1} = g^{ij}_{/1} = 0, \quad g_{ij|k} = g^{ij}_{|k} = 0, \quad g_{ij}|_{(k)}^{(1)} = g^{ij}|_{(k)}^{(1)} = 0.$$

For the study of the magnetized non-viscous plasma dynamics, in a generalized relativistic time-dependent Lagrangian geometrical approach on 1-jet spaces, we use the following geometrical objects (see also [41], [64]):

1. the unit relativistic time-dependent velocity-d-vector of a test particle, which is given by

$$U = u_1^i(t, x^k, y_1^k) \frac{\partial}{\partial y_1^i},$$

where, if we take $\varepsilon^2 = h^{11} g_{pq} y_1^p y_1^q > 0$, then we put $u_1^i = y_1^i / \varepsilon$. Obviously, we have $h^{11} u_{i1} u_1^i = 1$, where $u_{i1} = g_{im} u_1^m$;

2. the distinguished relativistic time-dependent 2-form of the (electric field)-(magnetic induction), which is given by

$$H = H_{ij}(t, x^k, y_1^k) dx^i \wedge dx^j;$$

3. the distinguished relativistic time-dependent 2-form of the (electric induction)-(magnetic field), which is given by

$$G = G_{ij}(t, x^k, y_1^k) dx^i \wedge dx^j;$$

4. the relativistic time-dependent Minkowski energy d-tensor of the electromagnetic field inside the non-isotropic plasma, which is given by

$$E = E_{ij}(t, x^k, y_1^k) dx^i \otimes dx^j + h^{11} E_{ij}(t, x^k, y_1^k) \delta y_1^i \otimes \delta y_1^j.$$

The adapted components of the relativistic time-dependent Minkowski energy are defined by

$$E_{ij} = \frac{1}{4} g_{ij} H_{rs} G^{rs} + g^{rs} H_{ir} G_{js},$$

where $G^{rs} = g^{rp} g^{sq} G_{pq}$. Moreover, we suppose that the adapted components of the relativistic time-dependent Minkowski energy verify the *non-isotropic Lorentz conditions* [64]

$$E_{i|m}^m u_1^i = 0, \qquad E_i^m|_{(m)}^{(1)} u_1^i = 0, \tag{7.26}$$

where $E_i^m = g^{mp}E_{pi}$. Obviously, if we use the notations $H_r^m = g^{mp}H_{pr}$ and $G_i^r = g^{rs}G_{si}$, we have

$$E_i^m = \frac{1}{4}\delta_i^m H_{rs}G^{rs} - H_r^m G_i^r,$$

where δ_i^m is the Kronecker symbol.

In our jet generalized Lagrangian geometrical approach, the relativistic time-dependent non-isotropic plasma is characterized by an energy-stress-momentum d-tensor $\mathcal{T}$, which is defined by [41], [64]

$$\mathcal{T} = T_{ij}(t, x^k, y_1^k)dx^i \otimes dx^j + h^{11}T_{ij}(t, x^k, y_1^k)\delta y_1^i \otimes \delta y_1^j,$$

where [32]

$$T_{ij} = \left(\rho + \frac{\mathbf{p}}{c^2}\right)h^{11}u_{i1}u_{j1} + \mathbf{p}g_{ij} + E_{ij}. \tag{7.27}$$

The entities $c = \text{constant}$, $\mathbf{p} = \mathbf{p}(t, x^k, y_1^k)$, and $\rho = \rho(t, x^k, y_1^k)$ have the physical meanings of speed of light, non-isotropic hydrostatic pressure, and non-isotropic proper mass density of plasma. Note that the adapted components of the energy-stress-momentum d-tensor $\mathcal{T}$, which characterizes the non-isotropic plasma, are

$$\mathcal{T}_{CF} = \begin{cases} T_{ij}, & \text{for} & C = i, \quad F = j, \\ h^{11}T_{ij}, & \text{for} & C = \overset{(1)}{(i)}, \quad F = \overset{(1)}{(j)}, \\ 0, & \text{otherwise.} \end{cases} \tag{7.28}$$

In the jet generalized Lagrange framework of plasma, we postulate that the following *non-isotropic conservation laws* of the components (7.27) and (7.28) are true:

$$\mathcal{T}_{A:M}^M = 0, \qquad \forall\, A \in \left\{1, i, \overset{(1)}{(i)}\right\}, \tag{7.29}$$

where the capital Latin letters $A, M, \ldots$ are indices of kind $1, i$ or $\overset{(1)}{(i)}$, "$_{:M}$" represents one of the local covariant derivatives $h_{\mathbb{R}}-, h_M-$, or $v-$, and

$$\mathcal{T}_A^M = \mathbb{G}^{MD}\mathcal{T}_{DA} = \begin{cases} T_i^m, & \text{for} & A = i, \quad M = m, \\ T_i^m, & \text{for} & A = \overset{(1)}{(i)}, \quad M = \overset{(1)}{(m)}, \\ 0, & \text{otherwise.} \end{cases}$$

Note that the d-tensor T_i^m is given by the formula

$$T_i^m = g^{mp}T_{pi} = \left(\rho + \frac{\mathbf{p}}{c^2}\right)h^{11}u_1^m u_{i1} + \mathbf{p}\delta_i^m + E_i^m.$$

It is easy to see that the jet non-isotropic conservation laws (7.29) reduce to the following local non-isotropic conservation equations:

$$T_{i|m}^m = 0, \qquad T_i^m\big|_{(m)}^{(1)} = 0. \tag{7.30}$$

Moreover, by direct computations, we deduce that the non-isotropic conservation equations (7.30) become

$$h^{11}\left[\left(\rho+\frac{\mathbf{P}}{c^2}\right)u_1^m\right]_{|m}u_{i1}+\left(\rho+\frac{\mathbf{P}}{c^2}\right)h^{11}u_1^m u_{i1|m}+\mathbf{P}_{,,i}-g_{ir}\overset{h}{\mathcal{F}}{}^r=0,$$

$$h^{11}\left[\left(\rho+\frac{\mathbf{P}}{c^2}\right)u_1^m\right]\Big|_{(m)}^{(1)}u_{i1}+\left(\rho+\frac{\mathbf{P}}{c^2}\right)h^{11}u_1^m u_{i1}\Big|_{(m)}^{(1)}$$

$$+\mathbf{P}_{\#(i)}^{(1)}-g_{ir}\overset{v}{\mathcal{F}}{}^{r1}=0,$$

(7.31)

where $\mathbf{P}_{,,i}=\delta\mathbf{p}/\delta x^i$, $\mathbf{p}_{\#(i)}^{(1)}=\partial\mathbf{p}/\partial y_1^i$, and

- $\overset{h}{\mathcal{F}}{}^r=-g^{rs}E_{s|m}^m$ is the *non-isotropic horizontal Lorentz force*;

- $\overset{v}{\mathcal{F}}{}^{r1}=-g^{rs}E_s^m\big|_{(m)}^{(1)}$ is the *non-isotropic vertical Lorentz d-tensor force*.

Contracting now the non-isotropic conservation equations (7.31) with u_1^i and taking into account the non-isotropic Lorentz conditions (7.26), we find the *non-isotropic continuity equations* of plasma, namely,

$$\left[\left(\rho+\frac{\mathbf{P}}{c^2}\right)u_1^m\right]_{|m}+\mathbf{P}_{,,m}u_1^m=0,$$

$$\left[\left(\rho+\frac{\mathbf{P}}{c^2}\right)u_1^m\right]\Big|_{(m)}^{(1)}+\mathbf{P}_{\#(m)}^{(1)}u_1^m=0,$$

(7.32)

where we also used the equalities

$$0=h^{11}u_{i1}u_1^i{}_{|m}=\frac{1}{2}\left(h^{11}u_{i1}u_1^i\right)_{,,m}=-h^{11}u_{i1|m}u_1^i,$$

$$0=h^{11}u_{i1}u_1^i\big|_{(m)}^{(1)}=\frac{1}{2}\left(h^{11}u_{i1}u_1^i\right)_{\#(m)}^{(1)}=-h^{11}u_{i1}\big|_{(m)}^{(1)}u_1^i,$$

the symbols "$_{,,m}$" and "$\overset{(1)}{\#(m)}$" being the derivative operators $\delta/\delta x^m$ and $\partial/\partial y_1^m$.

Replacing the continuity laws (7.32) into the conservation equations (7.31), we find the *non-isotropic relativistic Euler equations* for plasma, namely,

$$\left(\rho+\frac{\mathbf{P}}{c^2}\right)h^{11}u_{i1|m}u_1^m-\mathbf{P}_{,,m}\left(h^{11}u_1^m u_{i1}-\delta_i^m\right)-g_{im}\overset{h}{\mathcal{F}}{}^m=0,$$

$$\left(\rho+\frac{\mathbf{P}}{c^2}\right)h^{11}u_{i1}\big|_{(m)}^{(1)}u_1^m-\mathbf{P}_{\#(m)}^{(1)}\left(h^{11}u_1^m u_{i1}-\delta_i^m\right)-g_{im}\overset{v}{\mathcal{F}}{}^{m1}=0.$$

(7.33)

If we take now $y_1^m=dx^m/dt$, then we have

$$u_1^m=\frac{1}{\varepsilon_0}\frac{dx^m}{dt}=\frac{dx^m}{ds},\qquad \varepsilon_0^2=h^{11}(t)g_{ij}(t,x,dx/dt)\frac{dx^i}{dt}\frac{dx^j}{dt},$$

where s is a natural parameter of the curve $c = (x^k(t))$, having the geometrical property $ds/dt = \varepsilon_0$. Introducing this u_1^m into the non-isotropic Euler equations (7.33), we obtain the equations of the *non-isotropic stream lines* for jet plasma, which are given by the following ODE systems:

- *horizontal* non-isotropic stream line ODEs:

$$\frac{d^2 x^k}{ds^2} + \left[L_{rm}^k - \frac{c^2}{\mathbf{p} + \rho c^2} \delta_r^k \mathbf{P}_{,,m} \right] \frac{dx^r}{ds} \frac{dx^m}{ds} = \frac{h_{11}c^2}{\mathbf{p} + \rho c^2} \left[\overset{h}{\mathcal{F}}{}^k - g^{km} \mathbf{P}_{,,m} \right]$$

$$+ \frac{N_{(1)m}^{(k)}}{\varepsilon_0} \frac{dx^m}{ds} - \frac{h^{11} N_{(1)m}^{(p)} g_{pr}}{\varepsilon_0} \frac{dx^r}{ds} \frac{dx^m}{ds} \frac{dx^k}{ds}$$

$$- \frac{h^{11} N_{(1)m}^{(r)}}{2} \frac{\partial g_{pq}}{\partial y_1^r} \frac{dx^p}{ds} \frac{dx^q}{ds} \frac{dx^m}{ds} \frac{dx^k}{ds};$$

- *vertical* non-isotropic stream line ODEs:

$$\left[C_{r(m)}^{k(1)} - \frac{c^2}{\mathbf{p} + \rho c^2} \delta_r^k \mathbf{P}_{\#(m)}^{(1)} \right] \frac{dx^r}{ds} \frac{dx^m}{ds} = \frac{h_{11}c^2}{\mathbf{p} + \rho c^2} \left[\overset{v}{\mathcal{F}}{}^{k1} - g^{km} \mathbf{P}_{\#(m)}^{(1)} \right]$$

$$+ \frac{h^{11}}{2} \frac{\partial g_{pq}}{\partial y_1^r} \frac{dx^p}{ds} \frac{dx^q}{ds} \frac{dx^r}{ds} \frac{dx^k}{ds}.$$

Remark 141. If the metrical d-tensor $g_{ij}(t, x, y)$ is a Finslerian-like one, that is, we have

$$g_{ij}(t, x, y) = \frac{h_{11}}{2} \frac{\partial^2 F^2}{\partial y_1^i \partial y_1^j},$$

where $F : J^1(\mathbb{R}, M^n) \to \mathbb{R}_+$ is a jet Finslerian metric, then we use the *canonical spatial nonlinear connection* $N = \left(N_{(1)j}^{(k)} \right)$ of the jet Finsler space, whose general formula is given by (4.8) and (4.9). Consequently, the ODEs of the *stream lines* of plasma in non-isotropic jet Finsler spaces reduce to the following:

- *horizontal* non-isotropic stream line ODEs:

$$\frac{d^2 x^k}{ds^2} + \left[L_{rm}^k - \frac{c^2}{\mathbf{p} + \rho c^2} \delta_r^k \mathbf{P}_{,,m} \right] \frac{dx^r}{ds} \frac{dx^m}{ds} = \frac{h_{11}c^2}{\mathbf{p} + \rho c^2} \left[\overset{h}{\mathcal{F}}{}^k - g^{km} \mathbf{P}_{,,m} \right]$$

$$+ \frac{N_{(1)m}^{(k)}}{\varepsilon_0} \frac{dx^m}{ds} - \frac{h^{11} N_{(1)m}^{(p)} g_{pr}}{\varepsilon_0} \frac{dx^r}{ds} \frac{dx^m}{ds} \frac{dx^k}{ds};$$

- *vertical* non-isotropic stream line ODEs:

$$\mathbf{P}_{\#(m)}^{(1)} \left[h_{11} g^{mk} - \frac{dx^m}{ds} \frac{dx^k}{ds} \right] = h_{11} \overset{v}{\mathcal{F}}{}^{k1}, \tag{7.34}$$

where $\varepsilon_0 = F$ and, if the generalized Christoffel symbols of $g_{ij}(t, x, y)$ are

$$\Gamma^i_{jk}(t, x, y) = \frac{g^{im}}{2} \left(\frac{\partial g_{jm}}{\partial x^k} + \frac{\partial g_{km}}{\partial x^j} - \frac{\partial g_{jk}}{\partial x^m} \right),$$

then we have

$$N^{(k)}_{(1)l} y_1^l = \Gamma^k_{pq} y_1^p y_1^q + \frac{g^{km}}{2} \frac{\partial g_{mp}}{\partial t} y_1^p.$$

More particular, if we have a jet Minkowski-like metrical d-tensor $g_{ij} = g_{ij}(y)$, then the *horizontal* non-isotropic stream line ODEs of plasma simplify as

$$\frac{d^2 x^k}{ds^2} + \left[L^k_{rm} - \frac{c^2}{\mathbf{p} + \rho c^2} \delta^k_r \mathbf{P}_{,,m} \right] \frac{dx^r}{ds} \frac{dx^m}{ds}$$

$$= \frac{h_{11} c^2}{\mathbf{p} + \rho c^2} \left[\overset{h}{\mathcal{F}}^k - g^{km} \mathbf{P}_{,,m} \right]. \tag{7.35}$$

7.6.3   The non-isotropic plasma as a medium geometrized by the jet rheonomic Berwald-Moór metric of order four

As a particular case of the preceding geometrical-physics results, in this Subsection we work with a fixed spatial manifold of order four M^4, whose 1-jet space $J^1(\mathbb{R}, M^4)$ is endowed with the *jet relativistic rheonomic Berwald-Moór metric of order four*

$$\overset{\circ}{F}(t, y) = \sqrt{h^{11}(t)} \cdot \sqrt[4]{y_1^1 y_1^2 y_1^3 y_1^4}. \tag{7.36}$$

Recall that the canonical geometrical objects produced by the Berwald-Moór metric of order four (7.36) and the nonlinear connection (7.9) were computed in the preceding Sections. Also, note that, throughout this Subsection, the Latin letters run only from 1 to 4.

Consequently, the geometric dynamics of the non-isotropic plasma regarded as a medium geometrized by the jet rheonomic Berwald-Moór metric of order four (7.36) and the nonlinear connection (7.9) is obtained using the ODEs of stream lines (7.35) and (7.34) for the particular Berwald-Moór geometrical objects (7.6), (7.7), (7.10) and the relations (7.12) and (7.13). Therefore, we find the following geometrical equations for non-isotropic plasma:

- the ODEs of the *horizontal* Berwald-Moór non-isotropic stream lines of order four are given by

$$\frac{d^2 x^k}{ds^2} + \frac{c^2}{\mathbf{p} + \rho c^2} \mathbf{P}_{,,m} \left(1 - 4\delta^{km} \right) \frac{dx^m}{ds} \frac{dx^k}{ds} = \frac{h_{11} c^2}{\mathbf{p} + \rho c^2} \overset{h}{\mathcal{F}}^k,$$

where $k \in \{1, 2, 3, 4\}$ is a fixed index and we do sum by m;

- the ODEs of the *vertical* Berwald-Moór non-isotropic stream lines of order four are given by

$$\mathbf{P}^{(1)}_{\#(m)} \left(1 - 4\delta^{mk}\right) \frac{dx^m}{ds} \frac{dx^k}{ds} = h_{11} \overset{v}{\mathcal{F}}{}^{k1},$$

where $k \in \{1, 2, 3, 4\}$ is a fixed index and we do sum by m.

Remark 142. In the particular case when the hydrostatic pressure is dependent only by t and x [i.e., we have an isotropic hydrostatic pressure $\mathbf{p} = \mathbf{p}(t, x^k)$], the ODEs of stream lines for non-isotropic plasma endowed with Berwald-Moór metric of order four become:

- the ODEs of the *horizontal* Berwald-Moór non-isotropic stream lines of order four:

$$\frac{d^2 x^k}{ds^2} + \frac{c^2}{\mathbf{p} + \rho c^2} \mathbf{P}_{,m} \left(1 - 4\delta^{km}\right) \frac{dx^m}{ds} \frac{dx^k}{ds} = \frac{h_{11} c^2}{\mathbf{p} + \rho c^2} \overset{h}{\mathcal{F}}{}^k,$$

where $\mathbf{p}_{,m} = \partial \mathbf{p} / \partial x^m$;

- the ODEs of the *vertical* Berwald-Moór non-isotropic stream lines of order four:

$$\overset{v}{\mathcal{F}}{}^{k1} = 0.$$

Open Problem. Are there real physical interpretations for the geometric dynamics of non-isotropic plasma, produced by the jet rheonomic Berwald-Moór metric of order four and the nonlinear connection (7.9)?

CHAPTER 8

THE JET LOCAL SINGLE-TIME FINSLER-LAGRANGE GEOMETRY INDUCED BY THE RHEONOMIC CHERNOV METRIC OF ORDER FOUR

In the background of Finsler spaces which exhibit the total equivalence of all the non-isotropic directions and which are relevant for Relativity, an important model is given by the Chernov metric of order four ([28], [15]),

$$F : TM \to \mathbb{R}, \qquad F(y) = \sqrt[3]{y^1 y^2 y^3 + y^1 y^2 y^4 + y^1 y^3 y^4 + y^2 y^3 y^4}. \qquad (8.1)$$

The larger class of Finsler metrics to which this metric belongs (we refer to the m-root metrics) have been previously studied by the Japanese geometers Matsumoto and Shimada ([52], [53], [90]).

Considering the former geometrical and physical reasons, the present Chapter develops on the 1-jet space $J^1(\mathbb{R}, M^4)$ the Finsler-Lagrange geometry, together with its attached gravitational and electromagnetic field theory, for the natural rheonomic jet extension of the Chernov metric of order four,

$$F_{[3]}(t, y) = \sqrt{h^{11}(t)} \cdot \sqrt[3]{y_1^1 y_1^2 y_1^3 + y_1^1 y_1^2 y_1^4 + y_1^1 y_1^3 y_1^4 + y_1^2 y_1^3 y_1^4},$$

where $h_{11}(t)$ is a Riemannian metric on $\mathbb{R}$ and $(t, x^1, x^2, x^3, x^4, y_1^1, y_1^2, y_1^3, y_1^4)$ are the coordinates of the 1-jet space $J^1(\mathbb{R}, M^4)$. Consequently, in the sequel, we apply the general geometrical results from Chapter 4 to the jet rheonomic Chernov metric of order four $F_{[3]}$.

Jet Single-Time Lagrange Geometry and Its Applications **99**
1-st Edition. By Vladimir Balan and Mircea Neagu.
© 2011 John Wiley & Sons, Inc. Published 2011 John Wiley & Sons, Inc.

8.1 PRELIMINARY NOTATIONS AND FORMULAS

Let $(\mathbb{R}, h_{11}(t))$ be a Riemannian manifold, where $\mathbb{R}$ is the set of real numbers. The Christoffel symbol of the Riemannian metric $h_{11}(t)$ is

$$\kappa_{11}^1 = \frac{h^{11}}{2} \frac{dh_{11}}{dt}, \quad \text{where} \quad h^{11} = \frac{1}{h_{11}} > 0. \tag{8.2}$$

Let also M^4 be a real manifold of dimension four, whose local coordinates are (x^1, x^2, x^3, x^4). Let us consider the 1-jet space $J^1(\mathbb{R}, M^4)$, whose local coordinates are

$$(t, x^1, x^2, x^3, x^4, y_1^1, y_1^2, y_1^3, y_1^4).$$

These transform by the rules (the Einstein convention of summation is assumed)

$$\tilde{t} = \tilde{t}(t), \quad \tilde{x}^p = \tilde{x}^p(x^q), \quad \tilde{y}_1^p = \frac{\partial \tilde{x}^p}{\partial x^q} \frac{dt}{d\tilde{t}} \cdot y_1^q, \quad p, q = \overline{1,4}, \tag{8.3}$$

where $d\tilde{t}/dt \neq 0$ and rank $(\partial \tilde{x}^p / \partial x^q) = 4$.

We further consider that the manifold M^4 is endowed with a tensor of kind $(0,3)$, given by the local components $S_{pqr}(x)$, which is totally symmetric in the indices p, q, and r. We shall use the notations

$$S_{ij1} = 6S_{ijp}y_1^p, \quad S_{i11} = 3S_{ipq}(x)y_1^p y_1^q, \quad S_{111} = S_{pqr}y_1^p y_1^q y_1^r. \tag{8.4}$$

We assume that the d-tensor S_{ij1} is non-degenerate [i.e., there exists the d-tensor S^{jk1} on $J^1(\mathbb{R}, M^4)$, such that $S_{ij1}S^{jk1} = \delta_i^k$]. In this context, we can consider the *third-root Finsler-like function* [90], [12] (which is 1-positive homogenous in the variable y)

$$F(t, x, y) = \sqrt[3]{S_{pqr}(x)y_1^p y_1^q y_1^r} \cdot \sqrt{h^{11}(t)} = \sqrt[3]{S_{111}(x, y)} \cdot \sqrt{h^{11}(t)}, \tag{8.5}$$

where the Finsler function F has as the domain of definition all values (t, x, y) which satisfy the condition $S_{111}(x, y) \neq 0$. Then the 3-positive homogeneity of the "y-function" S_{111} [which is a d-tensor on the 1-jet space $J^1(\mathbb{R}, M^4)$] leads to the equalities

$$S_{i11} = \frac{\partial S_{111}}{\partial y_1^i}, \quad S_{i11}y_1^i = 3S_{111}, \quad S_{ij1}y_1^j = 2S_{i11}, \quad S_{ij1} = \frac{\partial S_{i11}}{\partial y_1^j} = \frac{\partial^2 S_{111}}{\partial y_1^i \partial y_1^j},$$

$$S_{ij1}y_1^i y_1^j = 6S_{111}, \quad \frac{\partial S_{ij1}}{\partial y_1^k} = 6S_{ijk}, \quad S_{ijp}y_1^p = \frac{1}{6}S_{ij1}.$$

The *fundamental metrical d-tensor* produced by F is given by the formula

$$g_{ij}(t, x, y) = \frac{h_{11}(t)}{2} \frac{\partial^2 F^2}{\partial y_1^i \partial y_1^j}.$$

By direct computations, the fundamental metrical d-tensor takes the form

$$g_{ij}(x, y) = \frac{S_{111}^{-1/3}}{3}\left[S_{ij1} - \frac{1}{3S_{111}}S_{i11}S_{j11}\right]. \tag{8.6}$$

Moreover, since the d-tensor S_{ij1} is non-degenerate, the matrix $g = (g_{ij})$ admits an inverse $g^{-1} = (g^{jk})$, whose entries are

$$g^{jk} = 3S_{111}^{1/3}\left[S^{jk1} + \frac{S_1^j S_1^k}{3\left(S_{111} - \mathfrak{S}_{111}\right)}\right], \tag{8.7}$$

where $S_1^j = S^{jp1}S_{p11}$ and $3\mathfrak{S}_{111} = S^{pq1}S_{p11}S_{q11}$.

8.2 THE RHEONOMIC CHERNOV METRIC OF ORDER FOUR

Beginning with this Section we will focus only on the *rheonomic Chernov metric of order four*, which is the Finsler-like metric (8.5) for the particular case

$$S_{pqr} := S_{[3]pqr} = \begin{cases} \dfrac{1}{3!}, & \{p, q, r\}\text{ - distinct indices,} \\ 0, & \text{otherwise.} \end{cases}$$

Consequently, the rheonomic Chernov metric of order four is given by

$$F_{[3]}(t, y) = \sqrt{h^{11}(t)} \cdot \sqrt[3]{y_1^1 y_1^2 y_1^3 + y_1^1 y_1^2 y_1^4 + y_1^1 y_1^3 y_1^4 + y_1^2 y_1^3 y_1^4}. \tag{8.8}$$

Moreover, using the preceding notations and formulas, we obtain the following relations:

$$S_{111} := S_{[3]111} = y_1^1 y_1^2 y_1^3 + y_1^1 y_1^2 y_1^4 + y_1^1 y_1^3 y_1^4 + y_1^2 y_1^3 y_1^4,$$

$$S_{i11} := S_{[3]i11} = \frac{\partial S_{[3]111}}{\partial y_1^i} = \frac{S_{[3]111}y_1^i - S_{[4]1111}}{\left(y_1^i\right)^2},$$

$$S_{ij1} := S_{[3]ij1} = \frac{\partial S_{[3]i11}}{\partial y_1^j} = \frac{\partial^2 S_{[3]111}}{\partial y_1^i \partial y_1^j} = \begin{cases} S_{[1]1} - y_1^i - y_1^j, & i \neq j, \\ 0, & i = j, \end{cases}$$

where $S_{[4]1111} = y_1^1 y_1^2 y_1^3 y_1^4$ and $S_{[1]1} = y_1^1 + y_1^2 + y_1^3 + y_1^4$. Note that, for $i \neq j$, the following equality holds true as well:

$$S_{[3]i11} \cdot S_{[3]j11} = S_{[3]111}\left(S_{[1]1} - y_1^i - y_1^j\right) + \frac{S_{[4]1111}^2}{\left(y_1^i\right)^2 \left(y_1^j\right)^2}.$$

Because we have $0 \neq \det \left(S_{ij1} \right)_{i,j=\overline{1,4}} = 4 \left[4S_{[4]1111} - S_{[1]1}S_{[3]111} \right] := \mathfrak{D}_{1111}$, we find

$$
S^{jk1} := S^{jk1}_{[3]} = \begin{cases} \dfrac{-2}{\mathfrak{D}_{1111}} \left(y_1^j + y_1^k \right) \left[y_1^j y_1^k + \dfrac{S_{[4]1111}}{y_1^j y_1^k} \right], & j \neq k, \\[3mm] \dfrac{1}{\mathfrak{D}_{1111}} \cdot \dfrac{1}{y_1^j} \left[\displaystyle\prod_{l=1}^{4} \left(y_1^j + y_1^l \right) \right], & j = k. \end{cases}
$$

Further, laborious computations lead to

$$
\begin{aligned}
S_1^j &:= S^j_{[3]1} = S^{jp1}_{[3]} S_{[3]p11} = \frac{1}{2} y_1^j, \\[2mm]
\mathfrak{S}_{111} &:= \mathfrak{S}_{[3]111} = S^{pq1}_{[3]} S_{[3]p11} S_{[3]q11} = \frac{1}{2} S_{[3]111}.
\end{aligned}
\tag{8.9}
$$

Replacing now the above computed entities into the formulas (8.6) and (8.7), we get $g_{ij} := g_{[3]ij} =$

$$
= \begin{cases} \dfrac{S^{-1/3}_{[3]111}}{9} \left[2 \left(S_{[1]1} - y_1^i - y_1^j \right) - \dfrac{S^2_{[4]1111}}{S_{[3]111}} \cdot \dfrac{1}{\left(y_1^i \right)^2 \left(y_1^j \right)^2} \right], & i \neq j, \\[4mm] -\dfrac{S^{-4/3}_{[3]111}}{9} \cdot S^2_{[3]i11}, & i = j, \end{cases}
\tag{8.10}
$$

and

$$
g^{jk} := g^{jk}_{[3]} = 3S^{1/3}_{[3]111} \left[S^{jk1} + \frac{1}{6S_{[3]111}} y_1^j y_1^k \right].
\tag{8.11}
$$

Consequently, using the general formulas (4.7), (4.8), and (4.9), we find the following geometrical result:

Proposition 143. *For the rheonomic Chernov metric of order four (8.8), the **energy action functional***

$$
\begin{aligned}
\mathbb{E}_{[3]}(t, x(t)) &= \int_a^b F^2_{[3]} \sqrt{h_{11}} \, dt \\
&= \int_a^b \sqrt[3]{\left(y_1^1 y_1^2 y_1^3 + y_1^1 y_1^2 y_1^4 + y_1^1 y_1^3 y_1^4 + y_1^2 y_1^3 y_1^4 \right)^2} \cdot h^{11} \sqrt{h_{11}} \, dt
\end{aligned}
$$

produces on the 1-jet space $J^1(\mathbb{R}, M^4)$ *the **canonical nonlinear connection***

$$
\Gamma = \left(M^{(i)}_{(1)1} = -\kappa^1_{11} y_1^i, \ N^{(i)}_{(1)j} = 0 \right).
\tag{8.12}
$$

Because the canonical nonlinear connection (8.12) has the spatial components equal to zero, it follows that our subsequent geometrical theory becomes trivial, in a way. For such a reason, in order to avoid the triviality of our theory and in order to have a certain kind of symmetry, we will use on the 1-jet space $J^1(\mathbb{R}, M^4)$, by an *a priori* definition, the following nonlinear connection (which does not curve the space):

$$\Gamma_{[3]} = \left(M_{(1)1}^{(i)} = -\kappa_{11}^1 y_1^i, \ N_{(1)j}^{(i)} = -\frac{\kappa_{11}^1}{2} \delta_j^i \right), \tag{8.13}$$

where δ_j^i is the Kronecker symbol.

Remark 144. The spatial components of the nonlinear connection (8.13), which are given on the local chart $\mathcal{U}$ by the functions

$$\mathring{N} = \left(N_{(1)j}^{(i)} = -\frac{\kappa_{11}^1}{2} \delta_j^i \right),$$

do not have a global character on the 1-jet space $J^1(\mathbb{R}, M^4)$, but have only a local character. Consequently, taking into account the transformation rules (1.8) and (1.18), it follows that the spatial nonlinear connection $\mathring{N}$ has in the local chart $\widetilde{\mathcal{U}}$ the following components:

$$\widetilde{N}_{(1)l}^{(k)} = -\frac{\widetilde{\kappa}_{11}^1}{2} \delta_l^k + \frac{1}{2} \frac{d\widetilde{t}}{dt} \frac{d^2 t}{d\widetilde{t}^2} \delta_l^k + \frac{\partial \widetilde{x}^k}{\partial x^m} \frac{\partial^2 x^m}{\partial \widetilde{x}^l \partial \widetilde{x}^r} \widetilde{y}_1^r.$$

8.3 CARTAN CANONICAL LINEAR CONNECTION. d-TORSIONS AND d-CURVATURES

The importance of the nonlinear connection (8.13) comes from the possibility of construction of the dual *local adapted bases* of d-vector fields

$$\left\{ \frac{\delta}{\delta t} = \frac{\partial}{\partial t} + \kappa_{11}^1 y_1^p \frac{\partial}{\partial y_1^p} \ ; \ \frac{\delta}{\delta x^i} = \frac{\partial}{\partial x^i} + \frac{\kappa_{11}^1}{2} \frac{\partial}{\partial y_1^i} \ ; \ \frac{\partial}{\partial y_1^i} \right\} \subset \mathcal{X}(E) \tag{8.14}$$

and d-covector fields

$$\left\{ dt \ ; \ dx^i \ ; \ \delta y_1^i = dy_1^i - \kappa_{11}^1 y_1^i dt - \frac{\kappa_{11}^1}{2} dx^i \right\} \subset \mathcal{X}^*(E), \tag{8.15}$$

where $E = J^1(\mathbb{R}, M^4)$. Note that, under a change of coordinates (8.3), the elements of the adapted bases (8.14) and (8.15) transform as classical tensors. Consequently, all subsequent geometrical objects on the 1-jet space $J^1(\mathbb{R}, M^4)$ (as Cartan canonical connection, torsion, curvature, etc.) will be described in local adapted components.

Using the general result of the Theorem 76, by direct computations, we can give the following important geometrical result:

Theorem 145. *The Cartan canonical $\Gamma_{[3]}$-linear connection, produced by the rheonomic Chernov metric of order four (8.8), has the following adapted local components:*

$$C\Gamma_{[3]} = \left(\kappa_{11}^1, \ G_{j1}^k = 0, \ L_{jk}^i = \frac{\kappa_{11}^1}{2} C_{j(k)}^{i(1)}, \ C_{j(k)}^{i(1)} \right),$$

where

$$C^{i(1)}_{j(k)} = 3S^{im1}_{[3]}S_{[3]jkm} + \frac{1}{9}\frac{1}{S^2_{[3]111}}S_{[3]j11}S_{[3]k11}y^i_1$$

$$-\frac{1}{6}\frac{1}{S_{[3]111}}\left[S_{[3]jk1}\frac{y^i_1}{2} + \delta^i_j S_{[3]k11} + \delta^i_k S_{[3]j11}\right]. \tag{8.16}$$

Proof: Using the Chernov derivative operators (8.14) and (8.15), together with the relations (8.10) and (8.11), we apply the general formulas which give the adapted components of the Cartan canonical connection (see Theorem 76):

$$G^k_{j1} = \frac{g^{km}_{[3]}}{2}\frac{\delta g_{[3]mj}}{\delta t}, \quad L^i_{jk} = \frac{g^{im}_{[3]}}{2}\left(\frac{\delta g_{[3]jm}}{\delta x^k} + \frac{\delta g_{[3]km}}{\delta x^j} - \frac{\delta g_{[3]jk}}{\delta x^m}\right),$$

$$C^{i(1)}_{j(k)} = \frac{g^{im}_{[3]}}{2}\left(\frac{\partial g_{[3]jm}}{\partial y^k_1} + \frac{\partial g_{[3]km}}{\partial y^j_1} - \frac{\partial g_{[3]jk}}{\partial y^m_1}\right) = \frac{g^{im}_{[3]}}{2}\frac{\partial g_{[3]jk}}{\partial y^m_1},$$

where, by computations, we have (for more details, see [90] and [12])

$$\frac{\partial g_{[3]jk}}{\partial y^m_1} = 2S^{-1/3}_{[3]111}S_{[3]jkm} + \frac{4}{27}S^{-7/3}_{[3]111}S_{[3]j11}S_{[3]k11}S_{[3]m11}$$

$$-\frac{1}{9}S^{-4/3}_{[3]111}\left\{S_{[3]jk1}S_{[3]m11} + S_{[3]km1}S_{[3]j11} + S_{[3]mj1}S_{[3]k11}\right\}.$$

Remark 146. The following properties of the d-tensor $C^{i(1)}_{j(k)}$ hold true:

$$C^{i(1)}_{j(k)} = C^{i(1)}_{k(j)}, \quad C^{i(1)}_{j(m)}y^m_1 = 0.$$

Theorem 147. *The Cartan canonical connection $C\Gamma_{[3]}$ of the rheonomic Chernov metric of order four (8.8) has* **three** *effective local torsion d-tensors:*

$$P^{(k)\,(1)}_{(1)i(j)} = -\frac{1}{2}\kappa^1_{11}C^{k(1)}_{i(j)}, \quad P^{k(1)}_{i(j)} = C^{k(1)}_{i(j)}, \quad R^{(k)}_{(1)1j} = \frac{1}{2}\left(\frac{d\kappa^1_{11}}{dt} - \kappa^1_{11}\kappa^1_{11}\right)\delta^k_j.$$

Proof: A general h-normal Γ-linear connection on the 1-jet space $J^1(\mathbb{R}, M^4)$ is characterized by *eight* effective d-tensors of torsion (see Theorem 55). For our Cartan canonical connection $C\Gamma_{[3]}$ these reduce to the following *three* (the other five cancel):

$$P^{(k)\,(1)}_{(1)i(j)} = \frac{\partial N^{(k)}_{(1)i}}{\partial y^j_1} - L^k_{ji}, \quad R^{(k)}_{(1)1j} = \frac{\delta M^{(k)}_{(1)1}}{\delta x^j} - \frac{\delta N^{(k)}_{(1)j}}{\delta t}, \quad P^{k(1)}_{i(j)} = C^{k(1)}_{i(j)}.$$

Theorem 148. *The Cartan canonical connection $C\Gamma_{[3]}$ of the rheonomic Chernov metric of order four (8.8) has* **three** *effective local curvature d-tensors:*

$$R^l_{ijk} = \frac{1}{4}\kappa^1_{11}\kappa^1_{11}S^{l(1)(1)}_{i(j)(k)}, \quad P^{l\,(1)}_{ij(k)} = \frac{1}{2}\kappa^1_{11}S^{l(1)(1)}_{i(j)(k)},$$

$$S^{l(1)(1)}_{i(j)(k)} = \frac{\partial C^{l(1)}_{i(j)}}{\partial y^k_1} - \frac{\partial C^{l(1)}_{i(k)}}{\partial y^j_1} + C^{m(1)}_{i(j)}C^{l(1)}_{m(k)} - C^{m(1)}_{i(k)}C^{l(1)}_{m(j)}.$$

Proof: A general h-normal Γ-linear connection on the 1-jet space $J^1(\mathbb{R}, M^4)$ is characterized by *five* effective d-tensors of curvature (see Theorem 58). For our Cartan canonical connection $C\Gamma_{[3]}$ these reduce to the following *three* (the other two cancel):

$$R^l_{ijk} = \frac{\delta L^l_{ij}}{\delta x^k} - \frac{\delta L^l_{ik}}{\delta x^j} + L^m_{ij} L^l_{mk} - L^m_{ik} L^l_{mj},$$

$$P^{l\ (1)}_{ij(k)} = \frac{\partial L^l_{ij}}{\partial y^k_1} - C^{l(1)}_{i(k)|j} + C^{l(1)}_{i(m)} P^{(m)\ (1)}_{(1)j(k)},$$

$$S^{l(1)(1)}_{i(j)(k)} = \frac{\partial C^{l(1)}_{i(j)}}{\partial y^k_1} - \frac{\partial C^{l(1)}_{i(k)}}{\partial y^j_1} + C^{m(1)}_{i(j)} C^{l(1)}_{m(k)} - C^{m(1)}_{i(k)} C^{l(1)}_{m(j)},$$

where

$$C^{l(1)}_{i(k)|j} = \frac{\delta C^{l(1)}_{i(k)}}{\delta x^j} + C^{m(1)}_{i(k)} L^l_{mj} - C^{l(1)}_{m(k)} L^m_{ij} - C^{l(1)}_{i(m)} L^m_{kj}.$$

Remark 149. We have denoted by " $_{/1}$," " $_{|i}$," and " $\Big|^{(1)}_{(i)}$" the Cartan covariant derivatives with respect to the corresponding $\mathbb{R}$–horizontal (temporal), M–horizontal, and vertical vector fields of the basis (8.14).

8.4 APPLICATIONS OF THE RHEONOMIC CHERNOV METRIC OF ORDER FOUR

8.4.1 Geometrical gravitational theory

From a physical point of view, on the 1-jet space $J^1(\mathbb{R}, M^4)$, the rheonomic Chernov metric of order four (8.8) produces the adapted metrical d-tensor

$$\mathbb{G}_{[3]} = h_{11} dt \otimes dt + g_{[3]ij} dx^i \otimes dx^j + h^{11} g_{[3]ij} \delta y^i_1 \otimes \delta y^j_1, \tag{8.17}$$

where $g_{[3]ij}$ is given by (8.10). This is regarded again as a "non-isotropic gravitational potential." Also, the nonlinear connection $\Gamma_{[3]}$ [given by (8.13) and used in the construction of the distinguished 1-forms δy^i_1] is regarded as a sort of "interaction" between (t)-, (x)-, and (y)-fields.

We postulate that the non-isotropic gravitational potential $\mathbb{G}_{[3]}$ is governed by the *geometrical Einstein equations*

$$\text{Ric}\left(C\Gamma_{[3]}\right) - \frac{\text{Sc}\left(C\Gamma_{[3]}\right)}{2} \mathbb{G}_{[3]} = \mathcal{K}\mathcal{T}, \tag{8.18}$$

where Ric $\left(C\Gamma_{[3]}\right)$ is the *Ricci d-tensor* associated to the Cartan canonical connection $C\Gamma_{[3]}$, Sc $\left(C\Gamma_{[3]}\right)$ is the *scalar curvature*, $\mathcal{K}$ is the *Einstein constant*, and $\mathcal{T}$ is the intrinsic *stress-energy d-tensor* of matter.

In this way, working with the adapted basis of vector fields (8.14), we can find the local geometrical Einstein equations for the rheonomic Chernov metric of order four (8.8). Firstly, by direct computations, we find the following

Theorem 150. *The Ricci d-tensor of the Cartan canonical connection* $C\Gamma_{[3]}$ *of the rheonomic Chernov metric of order four (8.8) has the following effective local Ricci d-tensor components:*

$$R_{ij} := R^r_{ijr} \qquad = \frac{1}{4}\kappa^1_{11}\kappa^1_{11}\mathbb{S}^{(1)(1)}_{(i)(j)},$$

$$P^{(1)}_{i(j)} = P^{(1)}_{(i)j} := P^{r\,(1)}_{ij(r)} \qquad = \frac{1}{2}\kappa^1_{11}\mathbb{S}^{(1)(1)}_{(i)(j)},$$

$$\mathbb{S}^{(1)(1)}_{(i)(j)} \qquad = -9S^{pq1}_{[3]}S^{rm1}_{[3]}\left(S_{[3]ijp}S_{[3]qrm} - S_{[3]ipr}S_{[3]jqm}\right)$$

$$+ \frac{1}{12}\frac{1}{S_{[3]111}}S_{[3]ij1} - \frac{1}{18}\frac{1}{S^2_{[3]111}}S_{[3]i11}S_{[3]j11},$$

where $\mathbb{S}^{(1)(1)}_{(i)(j)} = S^{r(1)(1)}_{i(j)(r)}$ *is the vertical Ricci d-tensor field.*

Proof: Using the equality (8.16), by laborious direct computations, we obtain the following equalities (we assume implicit summation by r and m):

$$\frac{\partial C^{r(1)}_{i(j)}}{\partial y^r_1} = 3\frac{\partial S^{rm1}_{[3]}}{\partial y^r_1}S_{[3]ijm} - \frac{1}{2}\frac{1}{S_{[3]111}}S_{[3]ij1} + \frac{5}{9}\frac{1}{S^2_{[3]111}}S_{[3]i11}S_{[3]j11},$$

$$\frac{\partial C^{r(1)}_{i(r)}}{\partial y^j_1} = 3\frac{\partial S^{rm1}_{[3]}}{\partial y^j_1}S_{[3]irm} - \frac{2}{3}\frac{1}{S_{[3]111}}S_{[3]ij1} + \frac{2}{3}\frac{1}{S^2_{[3]111}}S_{[3]i11}S_{[3]j11},$$

$$C^{m(1)}_{i(j)}C^{r(1)}_{m(r)} = 9S^{mp1}_{[3]}S^{rq1}_{[3]}S_{[3]ijp}S_{[3]mrq} + \frac{2}{9}\frac{1}{S^2_{[3]111}}S_{[3]i11}S_{[3]j11}$$

$$- \frac{1}{2}\frac{1}{S_{[3]111}}S^{rq1}_{[3]}\left\{S_{[3]irq}S_{[3]j11} + S_{[3]jrq}S_{[3]i11}\right\} - \frac{1}{6}\frac{1}{S_{[3]111}}S_{[3]ij1},$$

$$C^{m(1)}_{i(r)}C^{r(1)}_{m(j)} = 9S^{mp1}_{[3]}S^{rq1}_{[3]}S_{[3]irp}S_{[3]mjq} + \frac{1}{6}\frac{1}{S^2_{[3]111}}S_{[3]i11}S_{[3]j11}$$

$$- \frac{1}{2}\frac{1}{S_{[3]111}}S^{rq1}_{[3]}\left\{S_{[3]irq}S_{[3]j11} + S_{[3]jrq}S_{[3]i11}\right\} - \frac{1}{12}\frac{1}{S_{[3]111}}S_{[3]ij1}.$$

Finally, taking into account that we have

$$\mathbb{S}^{(1)(1)}_{(i)(j)} = S^{r(1)(1)}_{i(j)(r)} = \frac{\partial C^{r(1)}_{i(j)}}{\partial y^r_1} - \frac{\partial C^{r(1)}_{i(r)}}{\partial y^j_1} + C^{m(1)}_{i(j)}C^{r(1)}_{m(r)} - C^{m(1)}_{i(r)}C^{r(1)}_{m(j)},$$

and using the equalities

$$\frac{\partial S_{[3]}^{rm1}}{\partial y_1^r} S_{[3]ijm} = -6 S_{[3]}^{mp1} S_{[3]}^{rq1} S_{[3]ijp} S_{[3]mrq},$$

$$\frac{\partial S_{[3]}^{rm1}}{\partial y_1^j} S_{[3]irm} = -6 S_{[3]}^{mp1} S_{[3]}^{rq1} S_{[3]irp} S_{[3]jmq},$$

we obtain the required result.

Remark 151. The vertical Ricci d-tensor $\mathbb{S}_{(i)(j)}^{(1)(1)}$ has the following property of symmetry: $\mathbb{S}_{(i)(j)}^{(1)(1)} = \mathbb{S}_{(j)(i)}^{(1)(1)}$.

Proposition 152. *The scalar curvature of the Cartan canonical connection* $C\Gamma_{[3]}$ *of the rheonomic Chernov metric of order four (8.8) is given by*

$$Sc \left(C\Gamma_{[3]} \right) = \frac{4h_{11} + \kappa_{11}^1 \kappa_{11}^1}{4} \cdot \mathbb{S}^{11}, \quad where \quad \mathbb{S}^{11} = g_{[3]}^{pq} \mathbb{S}_{(p)(q)}^{(1)(1)}.$$

Proof: The general formula for the scalar curvature of a Cartan connection is (see Proposition 90)

$$Sc \left(C\Gamma_{[3]} \right) = g_{[3]}^{pq} R_{pq} + h_{11} g_{[3]}^{pq} \mathbb{S}_{(p)(q)}^{(1)(1)}.$$

Describing the global geometrical Einstein equations (8.18) in the adapted basis of vector fields (8.14), we find the following important geometrical and physical result:

Theorem 153. *The local* **geometrical Einstein equations** *that govern the non-isotropic gravitational potential (8.17), produced by the rheonomic Chernov metric of order four (8.8) and the nonlinear connection (8.13), are given by*

$$\begin{cases} \xi_{11} \mathbb{S}^{11} h_{11} = \mathcal{T}_{11}, \\[2mm] \dfrac{\kappa_{11}^1 \kappa_{11}^1}{4\mathcal{K}} \mathbb{S}_{(i)(j)}^{(1)(1)} + \xi_{11} \mathbb{S}^{11} g_{ij} = \mathcal{T}_{ij}, \\[2mm] \dfrac{1}{\mathcal{K}} \mathbb{S}_{(i)(j)}^{(1)(1)} + \xi_{11} \mathbb{S}^{11} h^{11} g_{ij} = \mathcal{T}_{(i)(j)}^{(1)(1)}, \end{cases} \quad (8.19)$$

$$\begin{cases} 0 = \mathcal{T}_{1i}, \quad 0 = \mathcal{T}_{i1}, \quad\quad 0 = \mathcal{T}_{(i)1}^{(1)}, \\[2mm] 0 = \mathcal{T}_{1(i)}^{(1)}, \quad \dfrac{\kappa_{11}^1}{2\mathcal{K}} \mathbb{S}_{(i)(j)}^{(1)(1)} = \mathcal{T}_{i(j)}^{(1)}, \quad \dfrac{\kappa_{11}^1}{2\mathcal{K}} \mathbb{S}_{(i)(j)}^{(1)(1)} = \mathcal{T}_{(i)j}^{(1)}, \end{cases} \quad (8.20)$$

where

$$\xi_{11} = -\frac{4h_{11} + \kappa_{11}^1 \kappa_{11}^1}{8\mathcal{K}}.$$

Remark 154. The local geometrical Einstein equations (8.19) and (8.20) impose as the stress-energy d-tensor of matter $\mathcal{T}$ to be symmetric. In other words, the stress-energy d-tensor of matter $\mathcal{T}$ must satisfy the local symmetry conditions

$$\mathcal{T}_{AB} = \mathcal{T}_{BA}, \quad \forall A, B \in \left\{ 1, \, i, \, \begin{matrix} (1) \\ (i) \end{matrix} \right\}.$$

8.4.2   Geometrical electromagnetic theory

For the rheonomic Chernov metric of order four (8.8) and the nonlinear connection (8.13), via the general formulas (4.10), we find the electromagnetic 2-form

$$\mathbb{F} := \mathbb{F}_{[3]} = 0.$$

In conclusion, the locally-Minkowski rheonomic Chernov geometrical electromagnetic theory of order four is trivial. In our opinion, this fact suggests that the metric (8.8) has rather gravitational connotations than electromagnetic ones in its t-deformed (x-independent) version, which leads to the need of considering x-dependent conformal deformations of the structure (as, e.g., recently proposed by Garas'ko in [36]).

CHAPTER 9

JET FINSLERIAN GEOMETRY OF THE CONFORMAL MINKOWSKI METRIC

This Chapter develops the Finsler-like geometry on the 1-jet space for the jet conformal Minkowski (JCM) metric, which naturally extends the Minkowski metric in the Chernov-Pavlov framework. To this aim, there are determined the canonical nonlinear connection, distinguished (d-) Cartan linear connection, d-torsions, and d-curvatures. The field geometrical gravitational and electromagnetic d-models based on the JCM metric are discussed.

9.1 INTRODUCTION

Let $(\mathbb{R}, h_{11}(t))$ be a Riemannian manifold, where $\mathbb{R}$ is the set of real numbers. The Christoffel symbol of the Riemannian metric $h_{11}(t)$ is

$$\kappa_{11}^1 = \frac{h^{11}}{2} \frac{dh_{11}}{dt}, \quad \text{where} \quad h^{11} = (h_{11})^{-1} > 0.$$

Let also M^4 be a manifold of dimension four, whose local coordinates are $x = (x^1, x^2, x^3, x^4)$. These manifolds produce the 1-jet space $J^1(\mathbb{R}, M^4)$, whose local coordinates are $(t; x; y)$, where $y = (y_1^1, y_1^2, y_1^3, y_1^4)$. We recall that these trans-

Jet Single-Time Lagrange Geometry and Its Applications
1-st Edition. By Vladimir Balan and Mircea Neagu.

form by the rules

$$\tilde{t} = \tilde{t}(t), \quad \tilde{x}^p = \tilde{x}^p(x^q), \quad \tilde{y}_1^p = \frac{\partial \tilde{x}^p}{\partial x^q} \frac{dt}{d\tilde{t}} \cdot y_1^q, \quad p, q = \overline{1,4}, \qquad (9.1)$$

where $d\tilde{t}/dt \neq 0$ and rank $(\partial \tilde{x}^p/\partial x^q) = 4$.

In recent physical and geometrical studies ([7], [37], [79], [80]), an important role is played by the jet Finsler-like metric

$$F_{[2]}(t, y) = \sqrt{h^{11}(t)} \cdot \sqrt{y_1^1 y_1^2 + y_1^1 y_1^3 + y_1^1 y_1^4 + y_1^2 y_1^3 + y_1^2 y_1^4 + y_1^3 y_1^4}, \qquad (9.2)$$

which produces the fundamental metrical d-tensor

$$g_{ij} := g_{[2]ij} = \frac{h_{11}(t)}{2} \frac{\partial^2 F_{[2]}^2}{\partial y_1^i \partial y_1^j} = \frac{1}{2}(1 - \delta_{ij}) \Rightarrow g^{jk} := g_{[2]}^{jk} = \frac{2}{3}(1 - 3\delta^{jk}).$$

Using again the *a priori* jet nonlinear connection

$$\Gamma_{[2]} = \left(M_{(1)1}^{(i)} = -\kappa_{11}^1 y_1^i, \ N_{(1)j}^{(i)} = -\frac{\kappa_{11}^1}{2} \delta_j^i \right), \qquad (9.3)$$

we deduce that the Cartan $\Gamma_{[2]}$-linear connection is

$$C\Gamma_{[2]} = \left(\kappa_{11}^1, \ G_{j1}^k = 0, \ L_{jk}^i = 0, \ C_{j(k)}^{i(1)} = 0 \right).$$

For the Cartan connection $C\Gamma_{[2]}$ all torsion d-tensors vanish, except

$$R_{(1)1j}^{(k)} = \frac{1}{2} \left[\frac{d\kappa_{11}^1}{dt} - \kappa_{11}^1 \kappa_{11}^1 \right] \delta_j^k,$$

and all curvature d-tensors are zero. Consequently, all Ricci d-tensors vanish and the scalar curvature cancels.

Consequently, the geometrical Einstein equations (8.18), produced by the jet Finslerian metric (9.2) and the nonlinear connection (9.3), become trivial, namely,

$$0 = \mathcal{T}_{AB}, \quad \forall \, A, B \in \left\{ 1, \ i, \ \genfrac{}{}{0pt}{}{(1)}{(i)} \right\}.$$

At the same time, the electromagnetic 2-form associated to the jet Finslerian metric (9.2) and the nonlinear connection (9.3) has the trivial form $\mathbb{F} := \mathbb{F}_{[2]} = 0$. In conclusion, both the metric-tensor based geometrical (gravitational and electromagnetic) theories are shown to be trivial for the case of the rheonomic jet locally Finslerian metric (9.2). Hence, for developing a non-trivial model, one may need to consider other closely related alternatives, such as the *jet conformal Minkowski (JCM) metric* $F : J^1(\mathbb{R}, M^4) \to \mathbb{R}$, defined by[4]

$$F(t, x, y) = e^{\sigma(x)} \cdot \sqrt{h^{11}(t)} \cdot \sqrt{y_1^1 y_1^2 + y_1^1 y_1^3 + y_1^1 y_1^4 + y_1^2 y_1^3 + y_1^2 y_1^4 + y_1^3 y_1^4}, \qquad (9.4)$$

---

[4]In the following we shall reduce the domain of the constructed geometric objects in order to ensure their existence and, where this is required, their smoothness.

where $\sigma(x)$ is a smooth non-constant function on M^4.

Remark 155. It is easy to verify that (as emphasized in the recent studies [28] and [80]) the geometrical object

$$G_{11}(y) \overset{def}{=} y_1^1 y_1^2 + y_1^1 y_1^3 + y_1^1 y_1^4 + y_1^2 y_1^3 + y_1^2 y_1^4 + y_1^3 y_1^4 \qquad (9.5)$$

is a quadratic form in $y = (y_1^1, y_1^2, y_1^3, y_1^4)$, whose canonical form is the Minkowski metric. Namely, denoting $x = (x^1, x^2, x^3, x^4)$, $\tilde{x} = (\tilde{x}^1, \tilde{x}^2, \tilde{x}^3, \tilde{x}^4)$, and

$$A = \begin{pmatrix} 1/\sqrt{6} & -1/\sqrt{3} & 1 & -1/\sqrt{6} \\ 1/\sqrt{6} & 2/\sqrt{3} & 0 & -1/\sqrt{6} \\ 1/\sqrt{6} & 0 & 0 & 3/\sqrt{6} \\ 1/\sqrt{6} & -1/\sqrt{3} & -1 & -1/\sqrt{6} \end{pmatrix},$$

if we apply on the product manifold $\mathbb{R} \times M^4$ the invertible linear coordinate transformation

$$t = \tilde{t}, \quad {}^T x = A \cdot {}^T \tilde{x},$$

then in the induced coordinates $(\tilde{t}, \tilde{x}, \tilde{y})$ on $J^1(\mathbb{R}, M^4)$, we have ${}^T y = A \cdot {}^T \tilde{y}$, and the JCM Finslerian metric (9.4) has the particular form

$$F(\tilde{t}, \tilde{x}, \tilde{y}) = e^{\sigma(\tilde{x} \cdot {}^T A)} \cdot \sqrt{h^{11}(\tilde{t})} \cdot \sqrt{(\tilde{y}_1^1)^2 - (\tilde{y}_1^2)^2 - (\tilde{y}_1^3)^2 - (\tilde{y}_1^4)^2}.$$

In the sequel, we apply the general geometrical results from Chapter 4 to the particular jet conformal Minkowski-metric (9.4).

9.2 THE CANONICAL NONLINEAR CONNECTION OF THE MODEL

Let's consider on $J^1(\mathbb{R}, M^4)$ the JCM metric (9.4), whose domain of definition consists of all values $(t; x; y)$ which satisfy the condition $G_{11}(y) > 0$, where G_{11} is given by (9.5). If we use the notation

$$S_{[1]1} = y_1^1 + y_1^2 + y_1^3 + y_1^4,$$

then the following relations are true:

$$G_{i1} \overset{def}{=} \frac{\partial G_{11}}{\partial y_1^i} = S_{[1]1} - y_1^i,$$

$$G_{ij} \overset{def}{=} \frac{\partial G_{i1}}{\partial y_1^j} = \frac{\partial^2 G_{11}}{\partial y_1^i \partial y_1^j} = 1 - \delta_{ij},$$

where δ_{ij} is the Kronecker symbol. Obviously, the homogeneity of degree 2 of the "y-function" G_{11} [which is in fact a d-tensor on $J^1(\mathbb{R}, M^4)$] leads to the equalities

$$G_{i1} y_1^i = 2G_{11}, \quad G_{ij} y_1^i y_1^j = 2G_{11}.$$

By direct computation, we get the following:

Lemma 156. (a) *The fundamental metrical d-tensor produced by the JCM Finslerian metric F is given by the formula*

$$g_{ij}(t, x, y) = \frac{h_{11}(t)}{2} \frac{\partial^2 F^2}{\partial y_1^i \partial y_1^j},$$

which in our case leads to

$$g_{ij}(x) = \frac{e^{2\sigma(x)}}{2} \left(1 - \delta_{ij}\right), \tag{9.6}$$

and the matrix $g = (g_{ij})$ admits the inverse $g^{-1} = (g^{jk})$, whose entries are

$$g^{jk}(x) = \frac{2e^{-2\sigma(x)}}{3} \left(1 - 3\delta^{jk}\right).$$

(b) *The divergence of the σ-diagonal vector field on M^4*

$$D_\sigma = \sigma(x)\frac{\partial}{\partial x^1} + \sigma(x)\frac{\partial}{\partial x^2} + \sigma(x)\frac{\partial}{\partial x^3} + \sigma(x)\frac{\partial}{\partial x^4}$$

has the expression

$$\operatorname{div} D_\sigma = \sigma_1 + \sigma_2 + \sigma_3 + \sigma_4,$$

where $\sigma_i = \partial\sigma/\partial x^i$.

Hence, using the general results from Chapter 4, we yield the following:

Proposition 157. *For the conformal JCM metric (9.4), the energy action functional*

$$
\begin{aligned}
\mathbb{E}(t, x(t)) &= \int_a^b F^2(t, x, y)\sqrt{h_{11}}\,dt = \int_a^b e^{2\sigma(x)} \cdot G_{11}(y) \cdot h^{11}(t)\sqrt{h_{11}(t)}\,dt \\
&= \int_a^b e^{2\sigma(x)} \left(y_1^1 y_1^2 + y_1^1 y_1^3 + y_1^1 y_1^4 + y_1^2 y_1^3 + y_1^2 y_1^4 + y_1^3 y_1^4\right) h^{11}\sqrt{h_{11}}\,dt,
\end{aligned}
$$

*where $y = dx/dt$, produces on the 1-jet space $J^1(\mathbb{R}, M^4)$ the **canonical nonlinear connection***

$$\Gamma = \left(M_{(1)1}^{(i)} = -\kappa_{11}^1 y_1^i, \ N_{(1)j}^{(i)}\right), \tag{9.7}$$

where

$$N_{(1)j}^{(i)} = \sigma_j y_1^i + \sigma_m y_1^m \delta_j^i + \left[\sigma_i - \frac{1}{3}\operatorname{div} D_\sigma\right]\left(S_{[1]1} - y_1^j\right).$$

Proof: The Euler-Lagrange equations of the energy action functional $\mathbb{E}$ are

$$\frac{d^2 x^i}{dt^2} + 2H_{(1)1}^{(i)}\left(t, x^k, y_1^k\right) + 2G_{(1)1}^{(i)}\left(t, x^k, y_1^k\right) = 0, \qquad y_1^k = \frac{dx^k}{dt}, \tag{9.8}$$

where we have the local geometrical components

$$
\begin{cases}
H^{(i)}_{(1)1} \overset{def}{=} -\dfrac{1}{2}\kappa^1_{11}(t)y^i_1, \\[2mm]
G^{(i)}_{(1)1} \overset{def}{=} \dfrac{h_{11}g^{ik}}{4}\left[\dfrac{\partial^2 F^2}{\partial x^m \partial y^k_1}y^m_1 - \dfrac{\partial F^2}{\partial x^k} + \dfrac{\partial^2 F^2}{\partial t \partial y^k_1}\right. \\[2mm]
\qquad\qquad \left. +\dfrac{\partial F^2}{\partial y^k_1}\kappa^1_{11}(t) + 2h^{11}\kappa^1_{11}g_{km}y^m_1\right] = \\[2mm]
\qquad = \sigma_m y^m_1 y^i_1 + \left[\sigma_i - \dfrac{1}{3}\operatorname{div} D_\sigma\right]\cdot G_{11},
\end{cases}
$$

which determine a *semispray* on the 1-jet space $J^1(\mathbb{R}, M^4)$. Its associated *canonical nonlinear connection* has the general form

$$
\Gamma = \left(M^{(i)}_{(1)1} = 2H^{(i)}_{(1)1} = -\kappa^1_{11}y^i_1, \quad N^{(i)}_{(1)j} = \dfrac{\partial G^{(i)}_{(1)1}}{\partial y^j_1}\right).
$$

9.3 CARTAN CANONICAL LINEAR CONNECTION. d-TORSIONS AND d-CURVATURES

The canonical nonlinear connection (9.7) is essential in constructing the dual *adapted bases* of distinguished (d-) vector fields

$$
\left\{\frac{\delta}{\delta t} = \frac{\partial}{\partial t} + \kappa^1_{11}y^p_1\frac{\partial}{\partial y^p_1}\;;\; \frac{\delta}{\delta x^i} = \frac{\partial}{\partial x^i} - N^{(p)}_{(1)i}\frac{\partial}{\partial y^p_1}\;;\; \frac{\partial}{\partial y^i_1}\right\} \subset \mathcal{X}(E) \qquad (9.9)
$$

and distinguished covector fields

$$
\left\{dt\;;\; dx^i\;;\; \delta y^i_1 = dy^i_1 - \kappa^1_{11}y^i_1 dt + N^{(i)}_{(1)p}dx^p\right\} \subset \mathcal{X}^*(E), \qquad (9.10)
$$

where $E = J^1(\mathbb{R}, M^4)$. Note that, under a change of coordinates (9.1), the elements of the adapted bases (9.9) and (9.10) transform as classical tensors. Consequently, all subsequent geometrical objects on the 1-jet space $J^1(\mathbb{R}, M^4)$, like Cartan canonical linear connection, torsion, curvature, etc., will be described in local adapted components.

In this respect, using a general result from Chapter 4, by direct computations, we have the following:

Proposition 158. *The Cartan canonical Γ-linear connection, produced by the jet conformal Minkowski metric (9.4), has the following adapted local components:*

$$
C\Gamma = \left(\kappa^1_{11}, \; G^k_{j1} = 0, \; L^i_{jk}, \; C^{i(1)}_{j(k)} = 0\right), \qquad (9.11)
$$

where

$$L^i_{jk} = \delta^i_j \sigma_k + \delta^i_k \sigma_j + (1 - \delta_{jk}) \sigma_i - \frac{1 - \delta_{jk}}{3} \, div \, D_\sigma.$$

Proof: Using the local derivative operators (9.9) and the general formulas which provide the adapted components of the Cartan canonical connection, we get

$$\begin{cases} G^k_{j1} = \dfrac{g^{km}}{2} \dfrac{\delta g_{mj}}{\delta t} = 0, \quad L^i_{jk} = \dfrac{g^{im}}{2} \left(\dfrac{\delta g_{jm}}{\delta x^k} + \dfrac{\delta g_{km}}{\delta x^j} - \dfrac{\delta g_{jk}}{\delta x^m} \right), \\[2mm] C^{i(1)}_{j(k)} = \dfrac{g^{im}}{2} \left(\dfrac{\partial g_{jm}}{\partial y^k_1} + \dfrac{\partial g_{km}}{\partial y^j_1} - \dfrac{\partial g_{jk}}{\partial y^m_1} \right) = 0. \end{cases}$$

Remark 159. It is straightforward to check the relation $L^i_{jk} = \dfrac{\partial N^{(i)}_{(1)j}}{\partial y^k_1}$, which, considering the homogeneity of degree 1 of the local functions $N^{(i)}_{(1)j}$, leads to

$$\frac{\partial N^{(i)}_{(1)j}}{\partial y^m_1} y^m_1 = N^{(i)}_{(1)j} \quad \Leftrightarrow \quad L^i_{jm} y^m_1 = N^{(i)}_{(1)j}. \tag{9.12}$$

Proposition 160. *The Cartan canonical Γ-linear connection $C\Gamma$ of the jet conformal Minkowski metric (9.4) has a **single** effective local torsion d-tensor, namely,*

$$R^{(l)}_{(1)jk} = \Re^l_{pjk} y^p_1,$$

where, using the notations

$$\sigma_{ij} = \frac{\partial^2 \sigma}{\partial x^i \partial x^j}, \quad grad \, \sigma = (\sigma_1, \sigma_2, \sigma_3, \sigma_4), \quad \| grad \, \sigma \|^2 = \sigma_1^2 + \sigma_2^2 + \sigma_3^2 + \sigma_4^2,$$

$$(div \, D_\sigma)_i = \frac{\partial (div \, D_\sigma)}{\partial x^i} = \sigma_{1i} + \sigma_{2i} + \sigma_{3i} + \sigma_{4i},$$

we have

$$\begin{aligned} \Re^l_{ijk} =\ & \delta^l_j (\sigma_{ik} - \sigma_i \sigma_k) - \delta^l_k (\sigma_{ij} - \sigma_i \sigma_j) \\ & + (1 - \delta_{ij})(\sigma_{lk} - \sigma_l \sigma_k) - (1 - \delta_{ik})(\sigma_{lj} - \sigma_l \sigma_j) \\ & + \frac{1}{3} (div \, D_\sigma)(\sigma_k - \sigma_j + \delta_{ik} \sigma_j - \delta_{ij} \sigma_k) \\ & + \left[\| grad \, \sigma \|^2 - \frac{1}{3} (div \, D_\sigma)^2 \right] (\delta^l_k - \delta^l_j + \delta_{ik} \delta^l_j - \delta_{ij} \delta^l_k) \\ & + \frac{1}{3} \left[(div \, D_\sigma)_j - (div \, D_\sigma)_k + \delta_{ij} (div \, D_\sigma)_k - \delta_{ik} (div \, D_\sigma)_j \right]. \end{aligned} \tag{9.13}$$

Proof: A general h-normal Γ-linear connection on the 1-jet space $J^1(\mathbb{R}, M^4)$ is characterized by *eight* effective d-tensors of torsion. For our Cartan canonical connection (9.11), these reduce only to *one* (the other seven cancel):

$$R^{(l)}_{(1)jk} = \frac{\delta N^{(l)}_{(1)j}}{\delta x^k} - \frac{\delta N^{(l)}_{(1)k}}{\delta x^j}.$$

Using now the expressions of the derivatives $\delta/\delta x^i$, formula (9.12), and the y-independence $L^i_{jk} = L^i_{jk}(x)$, we find

$$R^{(l)}_{(1)jk} = \mathfrak{R}^l_{pjk} y^p_1,$$

where

$$\mathfrak{R}^l_{ijk} := \frac{\partial L^l_{ij}}{\partial x^k} - \frac{\partial L^l_{ik}}{\partial x^j} + L^r_{ij} L^l_{rk} - L^r_{ik} L^l_{rj}.$$

Finally, laborious computations lead to expression (9.13) of the d-tensor $\mathfrak{R}^l_{ijk}$.

Proposition 161. *The Cartan canonical Γ-linear connection $C\Gamma$ of the jet conformal Minkowski metric (9.4) has a **single** effective local curvature d-tensor, namely,*

$$R^l_{ijk} = \mathfrak{R}^l_{ijk},$$

where $\mathfrak{R}^l_{ijk}$ is given by (9.13).

Proof: A general h-normal Γ-linear connection on the 1-jet space $J^1(\mathbb{R}, M^4)$ is characterized by *five* effective d-tensors of curvature. For our Cartan canonical connection (9.11), these reduce only to *one* (the other four cancel), namely,

$$
\begin{aligned}
R^l_{ijk} &\stackrel{def}{=} \frac{\delta L^l_{ij}}{\delta x^k} - \frac{\delta L^l_{ik}}{\delta x^j} + L^r_{ij} L^l_{rk} - L^r_{ik} L^l_{rj} + C^{l(1)}_{i(r)} R^{(r)}_{(1)jk} \\
&= \frac{\delta L^l_{ij}}{\delta x^k} - \frac{\delta L^l_{ik}}{\delta x^j} + L^r_{ij} L^l_{rk} - L^r_{ik} L^l_{rj} \\
&= \frac{\partial L^l_{ij}}{\partial x^k} - \frac{\partial L^l_{ik}}{\partial x^j} + L^r_{ij} L^l_{rk} - L^r_{ik} L^l_{rj} = \mathfrak{R}^l_{ijk}.
\end{aligned}
$$

9.4 GEOMETRICAL FIELD MODEL PRODUCED BY THE JET CONFORMAL MINKOWSKI METRIC

9.4.1 Gravitational-like geometrical model

From a geometric-physical point of view, on the 1-jet space $J^1(\mathbb{R}, M^4)$, the jet conformal Minkowski metric (9.4) produces the adapted metrical d-tensor

$$\mathbb{G} = h_{11} dt \otimes dt + g_{ij} dx^i \otimes dx^j + h^{11} g_{ij} \delta y^i_1 \otimes \delta y^j_1, \qquad (9.14)$$

where g_{ij} is given by (9.6). This may be regarded as a "non-isotropic gravitational potential." In such a "physical" context, the nonlinear connection Γ (used in the construction of the distinguished 1-forms δy^i_1) prescribes, probably, a kind of "interaction" between (t)-, (x)-, and (y)-fields.

We postulate that the non-isotropic gravitational potential $\mathbb{G}$ is governed by the *geometrical Einstein equations*

$$\text{Ric } (C\Gamma) - \frac{\text{Sc } (C\Gamma)}{2} \mathbb{G} = \mathcal{K}\mathcal{T}, \qquad (9.15)$$

where Ric $(C\Gamma)$ is the *Ricci d-tensor* associated to the Cartan canonical connection $C\Gamma$ (in Riemannian sense and using adapted bases), Sc $(C\Gamma)$ is the *scalar curvature*, $\mathcal{K}$ is the *Einstein constant*, and $\mathcal{T}$ is the intrinsic *stress-energy* d-tensor of matter [55].

In this way, working with the adapted basis of vector fields (9.9), we can find the local geometrical Einstein equations for the metric (9.4). First, by direct computations, we find the following:

Lemma 162. *The Ricci d-tensor of the Cartan canonical connection $C\Gamma$ of the metric (9.4) has a **single** effective local Ricci d-tensor, namely,*

$$
\begin{aligned}
R_{ij} &= -2\left(\sigma_{ij} - \sigma_i\sigma_j\right) \\
&+ \frac{1 - \delta_{ij}}{3}\left[3\Delta\sigma + 6\left\|\operatorname{grad}\sigma\right\|^2 - 2\left(\operatorname{div} D_\sigma\right)^2 - \mathfrak{S}\right],
\end{aligned}
\tag{9.16}
$$

where

$$
\Delta\sigma = \sigma_{11} + \sigma_{22} + \sigma_{33} + \sigma_{44}, \qquad \mathfrak{S} = \sum_{p,q=1}^{4}\sigma_{pq}.
$$

Proof: A general h-normal Γ-linear connection on the 1-jet space $J^1(\mathbb{R}, M^4)$ is characterized by *six* effective Ricci d-tensors. For our Cartan canonical connection (9.11), these reduce only to *one* (the other five cancel):

$$
R_{ij} \overset{def}{=} R^m_{ijm} = \mathfrak{R}^m_{ijm}.
$$

Then, a direct computation gives expression (9.16) of the Ricci d-tensor R_{ij}.

Lemma 163. *The scalar curvature Sc $(C\Gamma)$ of the Cartan canonical connection $C\Gamma$ of the jet conformal Minkowski metric (9.4) is given by*

$$
R = 4e^{-2\sigma}\left[3\Delta\sigma + 3\left\|\operatorname{grad}\sigma\right\|^2 - \left(\operatorname{div} D_\sigma\right)^2 - \mathfrak{S}\right].
\tag{9.17}
$$

Proof: The general formula of the scalar curvature of the Cartan connection yields

$$
\text{Sc }(C\Gamma) \overset{def}{=} g^{pq}R_{pq} := R,
$$

where R is given by (9.17).

By describing the global geometrical Einstein equations (9.15) in the adapted basis of vector fields (9.9), we find the following important geometrical and physical result:

Proposition 164. *The local **geometrical Einstein equations** that govern the non-isotropic gravitational potential $\mathbb{G}$ [produced by the jet conformal Minkowski metric (9.4)] are given by*

$$
R_{ij} - \frac{R}{2}g_{ij} = \mathcal{K}\mathcal{T}_{ij},
\tag{9.18}
$$

$$\begin{cases} -Rh_{11} = 2\mathcal{K}T_{11}, \quad 0 = T_{1i}, \quad 0 = T_{i1}, \quad 0 = T_{(i)1}^{(1)}, \\[2mm] 0 = T_{1(i)}^{(1)}, \qquad 0 = T_{i(j)}^{(1)}, \quad 0 = T_{(i)j}^{(1)}, \quad -Rh^{11}g_{ij} = 2\mathcal{K}T_{(i)(j)}^{(1)(1)}. \end{cases} \tag{9.19}$$

Remark 165. The Einstein geometrical equations (9.18) and (9.19) impose that the stress-energy d-tensor of matter $\mathcal{T}$ be symmetric. In other words, the stress-energy d-tensor of matter $\mathcal{T}$ must satisfy the local symmetry conditions

$$\mathcal{T}_{AB} = \mathcal{T}_{BA}, \quad \forall\, A, B \in \left\{ 1, i, \overset{(1)}{\scriptstyle(i)} \right\}.$$

Moreover, we must *a priori* have the equality $T_{(i)(j)}^{(1)(1)} h_{11} = g_{ij} T_{11} h^{11}$.

By direct computations, the geometrical Einstein equations (9.18) and (9.19) imply the following identities of the stress-energy d-tensor:[5]

$$T_1^1 \overset{def}{=} h^{11} T_{11} = -\frac{R}{2\mathcal{K}}, \quad T_1^m \overset{def}{=} g^{mr} T_{r1} = 0, \quad T_{(1)1}^{(m)} \overset{def}{=} h_{11} g^{mr} T_{(r)1}^{(1)} = 0,$$

$$T_i^1 \overset{def}{=} h^{11} T_{1i} = 0, \quad T_i^m \overset{def}{=} g^{mr} T_{ri} = \frac{1}{\mathcal{K}} \left(g^{mr} R_{ri} - \frac{R}{2}\delta_i^m \right),$$

$$T_{(i)}^{1(1)} \overset{def}{=} h^{11} T_{1(i)}^{(1)} = 0, \quad T_{(1)(i)}^{(m)(1)} \overset{def}{=} h_{11} g^{mr} T_{(r)(i)}^{(1)(1)} = -\frac{R}{2\mathcal{K}}\delta_i^m,$$

$$T_{(1)i}^{(m)} \overset{def}{=} h_{11} g^{mr} T_{(r)i}^{(1)} = 0, \quad T_{(i)}^{m(1)} \overset{def}{=} g^{mr} T_{r(i)}^{(1)} = 0. \tag{9.20}$$

Consequently, the following local identities for the stress-energy d-tensor of matter hold true:

$$\begin{cases} T_{1/1}^1 + T_{1|m}^m + T_{(1)1}^{(m)}|_{(m)}^{(1)} = T_{1/1}^1 = -\frac{1}{2\mathcal{K}}\frac{\delta R}{\delta t}, \\[2mm] T_{i/1}^1 + T_{i|m}^m + T_{(1)i}^{(m)}|_{(m)}^{(1)} = T_{i|m}^m = \frac{1}{\mathcal{K}} \left(g^{mr} R_{ri} - \frac{R}{2}\delta_i^m \right)_{|m}, \\[2mm] T_{(i)/1}^{1(1)} + T_{(i)|m}^{m(1)} + T_{(1)(i)}^{(m)(1)}|_{(m)}^{(1)} = T_{(1)(i)}^{(m)(1)}|_{(m)}^{(1)} = -\frac{1}{2\mathcal{K}}\frac{\partial R}{\partial y_1^i}, \end{cases} \tag{9.21}$$

where

[5] Summing over both indices m, r is assumed in (9.22), over r in (9.20), and over m in (9.21) and (9.23).

$$T^1_{1/1} \overset{def}{=} \frac{\delta T^1_1}{\delta t} + T^1_1 \kappa^1_{11} - T^1_1 \kappa^1_{11} = \frac{\delta T^1_1}{\delta t}, \quad T^m_{1|m} \overset{def}{=} \frac{\delta T^m_1}{\delta x^m} + T^r_1 L^m_{rm} = 0,$$

$$T^{(m)}_{(1)1}\big|^{(1)}_{(m)} \overset{def}{=} \frac{\partial T^{(m)}_{(1)1}}{\partial y^m_1} = 0, \quad T^1_{i/1} \overset{def}{=} \frac{\delta T^1_i}{\delta t} + T^1_i \kappa^1_{11} = 0,$$

$$T^m_{i|m} \overset{def}{=} \frac{\delta T^m_i}{\delta x^m} + T^r_i L^m_{rm} - T^m_r L^r_{im}, \quad T^{(m)}_{(1)i}\big|^{(1)}_{(m)} \overset{def}{=} \frac{\partial T^{(m)}_{(1)i}}{\partial y^m_1} = 0,$$

$$T^{1(1)}_{(i)/1} \overset{def}{=} \frac{\delta T^{1(1)}_{(i)}}{\delta t} + 2T^{1(1)}_{(i)} \kappa^1_{11} = 0, \quad T^{(m)(1)}_{(1)(i)}\big|^{(1)}_{(m)} \overset{def}{=} \frac{\partial T^{(m)(1)}_{(1)(i)}}{\partial y^m_1},$$

$$T^{m(1)}_{(i)|m} \overset{def}{=} \frac{\delta T^{m(1)}_{(i)}}{\delta x^m} + T^{r(1)}_{(i)} L^m_{rm} - T^{m(1)}_{(r)} L^r_{im} = 0. \tag{9.22}$$

Taking into account that we have the y-independence $R = R(x)$, we obtain the following result:

Corollary 166. *The stress-energy d-tensor of matter T must verify the following conservation geometrical laws:*

$$T^1_{1/1} = 0, \quad T^m_{i|m} = \frac{1}{K}\left[g^{mr}R_{ri} - \frac{R}{2}\delta^m_i\right]_{|m}, \quad T^{(m)(1)}_{(1)(i)}\big|^{(1)}_{(m)} = 0. \tag{9.23}$$

9.4.2 Related electromagnetic model considerations

On our particular 1-jet space $J^1(\mathbb{R}, M^4)$, the jet conformal Minkowski metric (9.4) (via the Lagrangian function $L = F^2$) produces the electromagnetic 2-form which, due to (9.12), trivially vanishes

$$\mathbb{F} = 0.$$

In conclusion, the jet conformal Minkowski extended electromagnetic geometrical model constructed on the 1-jet space $J^1(\mathbb{R}, M^4)$ is trivial. Namely, in our jet geometrical approach, the jet conformal Minkowski electromagnetism, produced only by the metric (9.4) alone, leads to null electromagnetic geometrical components and to tautological Maxwell-like equations. In our opinion, this fact suggests that the jet conformal geometrical structure (9.4) of the 1-jet space $J^1(\mathbb{R}, M^4)$ is suitable for modeling gravitation rather than electromagnetism.

PART II

APPLICATIONS OF THE JET SINGLE-TIME LAGRANGE GEOMETRY

CHAPTER 10

GEOMETRICAL OBJECTS PRODUCED BY A NONLINEAR ODEs SYSTEM OF FIRST-ORDER AND A PAIR OF RIEMANNIAN METRICS

10.1 HISTORICAL ASPECTS

According to Olver and Udrişte ([77], [94]), a lot of applicative problems arising from Physics [79], [80], Biology [68], [75] or Economy [95] can be modeled on 1-jet spaces. In such a context, we point out that the Riemann-Lagrange geometry of the 1-jet space $J^1(\mathbb{R}, M)$ (which was already developed in the preceding chapters) is a very applicative mathematical framework. For example, this geometry contains many fruitful ideas for the geometrical interpretation of the solutions of a given ODEs (or a PDEs system in [71]). In such a perspective, it is important to note that Udrişte proved in [94] that the orbits of a given vector field may be regarded as horizontal geodesics in a suitable Riemann-Lagrange geometrical structure, solving in this way an old open problem suggested by Poincaré [84], [85]: *Find the geometrical structure which transforms the field lines of a given vector field into geodesics.*

In the sequel, we present the main geometrical ideas used by Udrişte in order to solve the open problem of Poincaré (for more details, the reader is invited to consult the works [94] and [95]). In this direction, let us consider a Riemannian manifold $(M^n, \varphi_{ij}(x))$ and let us fix an arbitrary vector field $X = (X^i(x))$ on the manifold M. Obviously, the vector field X produces the first-order ODEs system (*the dynamical system)*

$$\frac{dx^i}{dt} = X^i(x(t)), \ \forall \, i = \overline{1,n}. \tag{10.1}$$

Differentiating the first-order ODEs system (10.1) and making a convenient arranging of the terms involved, in the Udrişte framework one constructs (via the vector field X, the Riemannian metric φ_{ij}, and its Christoffel symbols γ^i_{jk}) a second-order prolongation (*the single-time geometric dynamical system*) having the form

$$\frac{d^2x^i}{dt^2} + \gamma^i_{jk}\frac{dx^j}{dt}\frac{dx^k}{dt} = F^i_j\frac{dx^j}{dt} + \varphi^{ih}\varphi_{kj}\, X^j\nabla_h X^k, \ \forall \, i = \overline{1,n}, \tag{10.2}$$

where ∇ is the Levi-Civita connection of the Riemannian manifold (M, φ) and

$$F^i_j = \nabla_j X^i - \varphi^{ih}\varphi_{kj}\nabla_h X^k$$

characterizes the *helicity* of the vector field X.

Remark 167. Note that any solution of class C^2 of the first-order ODEs system (10.1) is also a solution for the second-order ODEs system (10.2). However, the converse is not always true.

The second-order ODEs system (10.2) is important because it is equivalent with the Euler-Lagrange equations of that so-called the *least squares Lagrangian function*

$$L_S : TM \to \mathbb{R}_+,$$

given by

$$L_S(x,y) = \frac{1}{2}\varphi_{ij}(x)\left(y^i - X^i(x)\right)\left(y^j - X^j(x)\right), \tag{10.3}$$

where (x^i, y^i) are the local coordinates on the tangent space TM.

It is obvious now that the field lines of class C^2 of the vector field X are the *global minimum points* of the *least squares energy action* attached to L_S, so these field lines are solutions of the Euler-Lagrange equations produced by L_S. Because the Euler-Lagrange equations of L_S are exactly equations (10.2), it follows that the solutions of class C^2 of the first-order ODEs system (10.1) are *horizontal geodesics* on the *Riemann-Lagrange manifold* [94]

$$(\mathbb{R} \times M, \ 1 + \varphi, \ N(^i_1)_j = \gamma^i_{jk}y^k - F^i_j).$$

Remark 168. The authors of this book believe that the preceding least squares variational method for the geometrical study of the ODEs system (10.1) can be reduced to a natural extension of the following well known and simple idea coming from linear algebra: **In any Euclidian vector space** $(V, \langle \, , \, \rangle)$ **the following equivalence holds true:**

$$v = 0_V \Leftrightarrow ||v||^2 = 0, \tag{10.4}$$

where $||v||^2 = \langle v, v \rangle$. This fact will be applied in the next Sections.

In order to find the jet geometrical objects that characterize a given ODEs system (which governs any applicative dynamical system), let us consider the jet fiber bundle of order one $J^1(\mathbb{R}, M) \to \mathbb{R} \times M$ associated to the Riemannian manifolds $(\mathbb{R}, h_{11}(t))$ and $(M, \varphi_{ij}(x^k))$. It is important to note that we use the notation $\mathcal{P} = (h_{11}, \varphi_{ij})$ for our pair of Riemannian metrics and the inverse tensors of the preceding Riemannian metrics are denoted by h^{11} and φ^{ij}. We recall that these tensors verify the formulas $h^{11} = 1/h_{11} > 0$ and $\sum_{m=1}^{n} \varphi^{im} \varphi_{mj} = \delta^i_j$. Also, we use the notations κ^1_{11} and γ^i_{jk} for the Christoffel symbols of the Riemannian metrics h_{11} and φ_{ij}.

Let us consider an unknown curve $c = (x^i(t))$ and an arbitrary given d-tensor field

$$X = \left(X^{(i)}_{(1)}(t, x^k) \right)$$

that define on the 1-jet space $J^1(\mathbb{R}, M)$ the following ODEs system of order one:

$$
\begin{aligned}
\frac{dx^i}{dt} &= X^{(i)}_{(1)}(t, x^k(t)), && \forall\, i = \overline{1, n} && \Leftrightarrow \\
y^i_1 &= X^{(i)}_{(1)}(t, x^k(t)), && \forall\, i = \overline{1, n} && \Leftrightarrow \\
y^i_1 &- X^{(i)}_{(1)}(t, x^k(t)) = 0, && \forall\, i = \overline{1, n},
\end{aligned}
\tag{10.5}
$$

where $y^i_1 = dx^i/dt$.

Remark 169. Recently, using the theory of Lie systems, Cariñena and de Lucas ([25]) intensively studied various geometrical features for a class of systems of first-order ordinary differential equations, whose general solutions can be described in terms of certain finite families of particular solutions and a set of constants (e.g., systems of first-order ordinary linear differential equations). The connection between their results and the further developed theory, is a subject of ongoing research for the authors of the present monograph.

10.2 SOLUTIONS OF ODEs SYSTEMS OF ORDER ONE AS HARMONIC CURVES ON 1-JET SPACES. CANONICAL NONLINEAR CONNECTIONS

In the sequel, let us show that the solutions of class C^2 of the ODEs system (10.5) can be naturally regarded as harmonic curves on 1-jet spaces. In this way, under our geometrical assumptions, we prove the following interesting qualitative-energetic result (see Theorem 98):

Theorem 170. *All solutions of class C^2 of the ODEs system (10.5) are harmonic curves on the 1-jet space $J^1(\mathbb{R}, M)$ for the time-dependent semispray*

$$\mathcal{S}_{ODEs}(\mathcal{P}) = \left(H^{(i)}_{(1)1}, G^{(i)}_{(1)1} \right),$$

whose components are given by the formulas

$$H_{(1)1}^{(i)} = -\frac{1}{2}\kappa_{11}^1 y_1^i, \quad G_{(1)1}^{(i)} = \frac{1}{2}\gamma_{jk}^i y_1^j y_1^k - h_{11}F^i(t, x^k, y_1^r),$$

where

$$F^i(t, x^k, y_1^r) = \frac{h^{11}}{2}\left\{\varphi^{il}X_{(1)\|l}^{(s)}\varphi_{sr}\left[X_{(1)}^{(r)} - y_1^r\right] + X_{(1)\|m}^{(i)}y_1^m + X_{(1)//1}^{(i)}\right\}$$

and

$$X_{(1)//1}^{(i)} = \frac{\partial X_{(1)}^{(i)}}{\partial t} - X_{(1)}^{(i)}\kappa_{11}^1, \quad X_{(1)\|j}^{(i)} = \frac{\partial X_{(1)}^{(i)}}{\partial x^j} + X_{(1)}^{(m)}\gamma_{mj}^i.$$

Proof: Using the idea from Linear Algebra (10.4) (which may be called in this case the *jet least squares variational method*), the initial first-order ODEs system (10.5) is equivalent with the ODEs system

$$\sum_{i,j=\overline{1,n}}\left\{h^{11}(t)\varphi_{ij}(x^k)\left(y_1^i - X_{(1)}^{(i)}(t, x^k(t))\right)\left(y_1^j - X_{(1)}^{(j)}(t, x^k(t))\right)\right\} = 0.$$

$$(10.6)$$

This is because the Riemannian metrics $h_{11}(t)$ and $\varphi_{ij}(x^k)$ are positive definite ones and they produce a distinguished vertical metric on $J^1(\mathbb{R}, M)$. It is obvious now that the local computations in the preceding ODEs system lead to the following new form of equations (10.6):

$$JLS_{\mathcal{P}}^{\text{ODEs}} := \sum_{i,j=1}^{n} h^{11}(t)\varphi_{ij}(x^k)y_1^i y_1^j + \sum_{1}^{n} U_{(i)}^{(1)}(t, x^k)y_1^i + \Phi(t, x^k) = 0, \quad (10.7)$$

where

$$U_{(1)}^{(i)}(t, x^k) = -2h^{11}\sum_{m=1}^{n}\varphi_{im}X_{(1)}^{(m)} \quad (10.8)$$

and

$$\Phi(t, x^k) = h^{11}\sum_{r,s=1}^{n}\varphi_{rs}X_{(1)}^{(r)}X_{(1)}^{(s)}. \quad (10.9)$$

Using the Einstein convention for summations, let us consider now the jet least squares time-dependent Lagrangian of electrodynamics type, which is given by

$$
\begin{aligned}
\mathcal{JLS}_{\mathcal{P}}^{\text{ODEs}} &:= JLS_{\mathcal{P}}^{\text{ODEs}}\sqrt{h_{11}(t)} = ||\mathbf{C} - \mathbf{X}||^2\sqrt{h_{11}(t)} \\
&= \left\{h^{11}(t)\varphi_{ij}(x^k)\left[y_1^i - X_{(1)}^{(i)}\right]\left[y_1^j - X_{(1)}^{(j)}\right]\right\}\sqrt{h_{11}(t)} \\
&= \left\{h^{11}(t)\varphi_{ij}(x^k)y_1^i y_1^j + U_{(i)}^{(1)}(t, x^k)y_1^i + \Phi(t, x^k)\right\}\sqrt{h_{11}(t)},
\end{aligned}
$$

where $\mathbf{C} = y_1^i(\partial/\partial y_1^i)$ and $\mathbf{X} = X_{(1)}^{(i)}(\partial/\partial y_1^i)$. Let us also consider the *jet least squares energy action functional*

$$\mathbb{E}_{\mathcal{JLS}_{\mathcal{P}}^{\text{ODEs}}} : C^2([a, b], M) \to \mathbb{R}_+,$$

given by

$$\mathbb{E}_{\mathcal{JLS}_{\mathcal{P}}^{\text{ODEs}}}(c) = \int_a^b \mathcal{JLS}_{\mathcal{P}}^{\text{ODEs}} dt = \int_a^b ||\mathbf{C} - \mathbf{X}||^2 \sqrt{h_{11}(t)} dt \geq 0.$$

It is obvious now that a smooth curve $c \in C^2([a, b], M)$, locally expressed by $c(t) = (x^1(t), x^2(t), ..., x^n(t))$, is a solution of the ODEs system (10.7) if and only if the curve c cancels the time-dependent Lagrangian $\mathcal{JLS}_{\mathcal{P}}^{\text{ODEs}}$. In other words, the curve c is a solution of the ODEs system (10.7) if and only if c is a global minimum point for the jet least squares energy action functional $\mathbb{E}_{\mathcal{JLS}_{\mathcal{P}}^{\text{ODEs}}}$. Therefore, every curve $c = (x^i(t))$ of class C^2 is a solution of the initial ODEs system (10.5) if and only if it verifies the Euler-Lagrange equations

$$\frac{\partial[\mathcal{JLS}_{\mathcal{P}}^{\text{ODEs}}]}{\partial x^i} - \frac{d}{dt}\left(\frac{\partial[\mathcal{JLS}_{\mathcal{P}}^{\text{ODEs}}]}{\partial y_1^i}\right) = 0, \quad \forall\, i = \overline{1, n}. \tag{10.10}$$

Taking into account the expression of the time-dependent least squares Lagrangian $\mathcal{JLS}_{\mathcal{P}}^{\text{ODEs}}$ and some local differential computations in the Euler-Lagrange equations (10.10), we claim (cf. Theorem 98) that equations (10.10) can be rewritten in the form (1.12) of the second-order ODEs system of the harmonic curves of the time-dependent semispray

$$\mathcal{S}_{\text{ODEs}}(\mathcal{P}) = \left(H_{(1)1}^{(i)}, G_{(1)1}^{(i)}\right), \tag{10.11}$$

whose temporal components are given by

$$H_{(1)1}^{(i)} = -\frac{1}{2}\kappa_{11}^1 y_1^i$$

and the spatial components are expressed by

$$G_{(1)1}^{(i)} = \frac{1}{2}\gamma_{jk}^i y_1^j y_1^k + \frac{h_{11}\varphi^{il}}{4}\left[U_{(l)m}^{(1)} y_1^m + \frac{\partial U_{(l)}^{(1)}}{\partial t} + U_{(l)}^{(1)}\kappa_{11}^1 - \frac{\partial \Phi}{\partial x^l}\right],$$

where

$$U_{(i)j}^{(1)} = \frac{\partial U_{(i)}^{(1)}}{\partial x^j} - \frac{\partial U_{(j)}^{(1)}}{\partial x^i}.$$

In what follows, using the expressions which give $U_{(i)}^{(1)}$ and Φ [see formulas (10.8) and (10.9)], together with some direct local computations, we find

$$U_{(i)j}^{(1)} = -2h^{11}\left[\varphi_{im}X_{(1)\|j}^{(m)} - \varphi_{jm}X_{(1)\|i}^{(m)}\right],$$

$$\frac{\partial U_{(i)}^{(1)}}{\partial t} + U_{(i)}^{(1)}\kappa_{11}^1 = -2h^{11}\varphi_{im}X_{(1)//1}^{(m)},$$

$$\frac{\partial\Phi}{\partial x^l} = 2h^{11}\varphi_{mr}X_{(1)}^{(m)}X_{(1)\|l}^{(r)},$$

where

$$X_{(1)//1}^{(i)} = \frac{\partial X_{(1)}^{(i)}}{\partial t} - X_{(1)}^{(i)}\kappa_{11}^1, \quad X_{(1)\|j}^{(i)} = \frac{\partial X_{(1)}^{(i)}}{\partial x^j} + X_{(1)}^{(m)}\gamma_{mj}^i \qquad (10.12)$$

are the local covariant derivatives of the Berwald $\overset{\circ}{\Gamma}$-linear connection $B\overset{\circ}{\Gamma}$ produced by the pair of Riemannian metrics $\mathcal{P} = (h_{11}, \varphi_{ij})$.

In conclusion, all our preceding computations imply the required result.

Definition 171. The time-dependent semispray

$$\mathcal{S}_{\text{ODEs}}(\mathcal{P}) = \left(H_{(1)1}^{(i)}, G_{(1)1}^{(i)}\right)$$

given by Theorem 170 is called the *canonical relativistic time-depedent semispray produced by the first-order ODEs system (10.5) and the pair of Riemannian metrics* $\mathcal{P} = (h_{11}, \varphi_{ij})$.

Remark 172. In other words, Theorem 170 asserts that the C^2 solutions of the initial first-order ODEs system (10.5) are verifying the second-order ODEs system (1.12) of the harmonic curves for the time-dependent semispray $\mathcal{S}_{\text{ODEs}}(\mathcal{P})$.

Remark 173. It is well known that, in Theoretical Physics, the Riemannian metrics are modeling the *gravitational potentials* of a space of events [92]. Thus, by a natural extension, we may assert that, in our context, the Riemannian metrics may model some *gravitational potentials of the abstract microscopic phenomena* intrinsically produced by the matter governed by the given ODEs system (10.5).

Taking into account the geometrical connection between the time-dependent semisprays and the nonlinear connections on 1-jet spaces (given by Proposition 26), we easily deduce the following important geometrical Corollary and Definition:

Corollary 174. *On the 1-jet space* $J^1(\mathbb{R}, M)$, *the* **canonical nonlinear connection**

$$\Gamma_{\mathcal{S}_{\text{ODEs}}(\mathcal{P})} = \left(M_{(1)1}^{(i)}, N_{(1)j}^{(i)}\right),$$

which is **produced by the ODEs system (10.5) and the pair of Riemannian metrics** $\mathcal{P} = (h_{11}, \varphi_{ij})$, *has the components*

$$M_{(1)1}^{(i)} = -\kappa_{11}^1 y_1^i \quad and \quad N_{(1)j}^{(i)} = \gamma_{jk}^i y_1^k - \mathcal{H}_{(1)j}^{(i)}, \qquad (10.13)$$

where

$$\mathcal{H}_{(1)j}^{(i)} = \frac{1}{2} \left[X_{(1)||j}^{(i)} - \varphi^{ir} X_{(1)||r}^{(s)} \varphi_{sj} \right].$$

10.3 FROM FIRST-ORDER ODEs SYSTEMS AND RIEMANNIAN METRICS TO GEOMETRICAL OBJECTS ON 1-JET SPACES

The geometrization on the 1-jet fiber bundle $J^1(\mathbb{R}, M)$ of the relativistic time-dependent Lagrangians was completely developed in Chapter 5: "The jet single-time electrodynamics." Moreover, some generalized geometrical gravitational and electromagnetic field theories were described over there, too.

Consequently, in this Section, we will particularize the main geometrical and physical results from Chapter 5 for our particular time-dependent Lagrangian of electrodynamics type $\mathcal{JLS}_{\mathcal{P}}^{\text{ODEs}}$. We underline that the jet least squares Lagrangian $\mathcal{JLS}_{\mathcal{P}}^{\text{ODEs}}$ from Theorem 170 is derived only from the given ODEs system (10.5) and the pair of Riemannian metrics $\mathcal{P} = (h_{11}, \varphi_{ij})$. As a consequence, we introduce the following concept:

Definition 175. The pair

$$\mathcal{JLS}_{\mathcal{P}}^{\text{ODEs}} RRL_1^n = \left(J^1(\mathbb{R}, M), \ JLS_{\mathcal{P}}^{\text{ODEs}} \right),$$

endowed with the nonlinear connection $\Gamma_{\mathcal{S}_{\text{ODEs}}(\mathcal{P})}$ given by (10.13), is called the *canonical relativistic time-dependent Lagrange space produced by the ODEs system (10.5) and the pair of Riemannian metrics* $\mathcal{P} = (h_{11}, \varphi_{ij})$.

In this geometrical context, using general formulas from Chapter 5, together with local computations, we find the following important differential geometric results (see Theorem 101):

Theorem 176. (i) *The* **canonical generalized Cartan connection** $C\Gamma_{\mathcal{S}_{\text{ODEs}}(\mathcal{P})}$ *of the relativistic time-dependent Lagrange space* $\mathcal{JLS}_{\mathcal{P}}^{\text{ODEs}} RRL_1^n$ *has the adapted components*

$$C\Gamma_{\mathcal{S}_{\text{ODEs}}(\mathcal{P})} = (\kappa_{11}^1, 0, \gamma_{jk}^i, 0).$$

(ii) *The* **torsion** d-tensor $\mathbf{T}$ *of the canonical Cartan connection* $C\Gamma_{\mathcal{S}_{\text{ODEs}}(\mathcal{P})}$ *of the space* $\mathcal{JLS}_{\mathcal{P}}^{\text{ODEs}} RRL_1^n$ *is determined by* **two** *adapted local d-tensors:*

$$R_{(1)1j}^{(i)} = \frac{1}{2} \left[X_{(1)||j//1}^{(i)} - \varphi^{ir} X_{(1)||r//1}^{(s)} \varphi_{sj} \right],$$

$$R_{(1)jk}^{(i)} = \mathfrak{R}_{jkm}^i y_1^m - \frac{1}{2} \left[X_{(1)||j||k}^{(i)} - \varphi^{ir} X_{(1)||r||k}^{(s)} \varphi_{sj} \right],$$

where $\mathfrak{R}_{ijk}^l$ *are the local curvature tensors of Riemannian metric* φ_{ij} *and the second covariant derivatives of the d-tensor* $X_{(1)}^{(i)}(t, x^k)$ *are given by*

$$X_{(1)||j//1}^{(i)} = \frac{\partial X_{(1)||j}^{(i)}}{\partial t} - X_{(1)||j}^{(i)} \kappa_{11}^1,$$

$$X^{(i)}_{(1)||j||k} = \frac{\partial X^{(i)}_{(1)||j}}{\partial x^k} + X^{(m)}_{(1)||j}\gamma^i_{mk} - X^{(i)}_{(1)||m}\gamma^m_{jk}.$$

(iii) *The **curvature** d-tensor* $\mathbf{R}$ *of the canonical generalized Cartan connection* $C\Gamma_{\mathcal{S}_{ODEs}(\mathcal{P})}$ *of the space* $\mathcal{JLS}^{ODEs}_{\mathcal{P}}RRL^n_1$ *is determined only by **one** adapted component, namely,*

$$R^l_{ijk} = \mathfrak{R}^l_{ijk},$$

that is exactly the components of the classical curvature tensor of the Riemannian metric φ_{ij}.

In the sequel, following the geometrical and abstract physical ideas from Chapter 5, by direct local computations, we can construct the electromagnetic distinguished 2-form of the space $\mathcal{JLS}^{ODEs}_{\mathcal{P}}RRL^n_1$ and describe its geometrical Maxwell equations [see formulas (5.2) and (5.3)].

Theorem 177. (i) *The **electromagnetic field** $\mathbb{F}$ of the relativistic time-dependent Lagrange space* $\mathcal{JLS}^{ODEs}_{\mathcal{P}}RRL^n_1$ *is expressed by the distinguished 2-form*

$$\mathbb{F} = F^{(1)}_{(i)j}\delta y^i_1 \wedge dx^j,$$

where

$$\delta y^i_1 = dy^i_1 + M^{(i)}_{(1)1}dt + N^{(i)}_{(1)j}dx^j$$

and

$$F^{(1)}_{(i)j} = \frac{h^{11}}{2}\left[\varphi_{im}X^{(m)}_{(1)||j} - \varphi_{jm}X^{(m)}_{(1)||i}\right].$$

(ii) *The adapted components* $F^{(1)}_{(i)j}$ *of the electromagnetic field* $\mathbb{F}$ *are governed by the following **geometrical Maxwell equations***

$$\begin{cases} F^{(1)}_{(i)j//1} = \dfrac{1}{4}\mathcal{A}_{\{i,j\}}\left\{h^{11}\varphi_{im}\left[X^{(m)}_{(1)||j//1} - \varphi^{mr}X^{(s)}_{(1)||r//1}\varphi_{sj}\right]\right\}, \\[2mm] \sum_{\{i,j,k\}} F^{(1)}_{(i)j||k} = 0, \end{cases}$$

where $\mathcal{A}_{\{i,j\}}$ *represents an alternate sum and* $\sum_{\{i,j,k\}}$ *means a cyclic sum.*

Remark 178. We do not describe the jet single-time gravitational theory of the relativistic time-dependent Lagrange space $\mathcal{JLS}^{ODEs}_{\mathcal{P}}RRL^n_1$ because its time-dependent gravitational field $\mathbb{G}$ and its attached geometrical Einstein equations are independent on the distinguished tensorial components $X^{(i)}_{(1)}(t, x^k)$ that define the ODEs system (10.5). In other words, the geometric time-dependent gravitational entities are depending only by the pair of Riemannian metrics $\mathcal{P} = (h_{11}, \varphi_{ij})$ (see Section 5.3). In conclusion, the jet single-time gravitational theory of the space $\mathcal{JLS}^{ODEs}_{\mathcal{P}}RRL^n_1$ does not contain informations provided by the ODEs system (10.5).

10.4 GEOMETRICAL OBJECTS PRODUCED ON 1-JET SPACES BY FIRST-ORDER ODEs SYSTEMS AND PAIRS OF EUCLIDIAN METRICS. JET YANG-MILLS ENERGY

In order to use the preceding geometrical results for the study of some ODEs systems of order one coming from diverse Applied Sciences, let us consider the particular Euclidian metrics as the pair $\mathcal{P}$ of Riemannian metrics:

$$\mathcal{P} := \Delta = \left(h_{11} = 1, \ \varphi_{ij} = \delta_{ij} \right),$$

where

$$\delta_{ij} = \left\{ \begin{array}{ll} 0, & i \neq j, \\ 1, & i = j, \end{array} \right.$$

is the classical Kronecker symbol.

In this particular situation, we are placed on the 1-jet space $J^1(\mathbb{R}, \mathbb{R}^n)$ produced by the Euclidian manifolds $(\mathbb{R}, 1)$ and $(\mathbb{R}^n, \delta_{ij})$. Consequently, Theorem 170 asserts now that all solutions of C^2-class of the ODEs system (10.5) may be regarded as the harmonic curves of the relativistic time-dependent semispray

$$\mathcal{S}_{\mathrm{ODEs}}(\Delta) = \left(H_{(1)1}^{(i)}, G_{(1)1}^{(i)} \right),$$

whose components are given by the formulas

$$H_{(1)1}^{(i)} = 0, \quad G_{(1)1}^{(i)} = -\frac{1}{2} \left\{ X_{(1)\|i}^{(r)} \left[X_{(1)}^{(r)} - y_1^r \right] + X_{(1)\|m}^{(i)} y_1^m + X_{(1)//1}^{(i)} \right\},$$

where

$$X_{(1)//1}^{(i)} = \frac{\partial X_{(1)}^{(i)}}{\partial t} \quad \text{and} \quad X_{(1)\|j}^{(i)} = \frac{\partial X_{(1)}^{(i)}}{\partial x^j}.$$

In other words, following the Proof of Theorem 170, we conclude that the solutions of C^2-class of the ODEs system (10.5) are minimizing the jet least squares time-dependent Lagrangian

$$\mathcal{LSQ}^{\mathrm{ODEs}}(t, x^k, y_1^k) = \sum_{i=1}^{n} \left(y_1^i - X_{(1)}^{(i)}(t, x^k) \right)^2,$$

which is obviously the time-dependent Lagrangian of electrodynamics type $\mathcal{J}\mathcal{LS}_{\mathcal{P}}^{\mathrm{ODEs}}$ for the particular pair of Euclidian metrics $\Delta = (1, \delta_{ij})$.

We shall further denote the relativistic time-dependent Lagrange space produced by the ODEs system (10.5) and the pair of Euclidian metrics $\Delta = (1, \delta_{ij})$ by

$$\mathcal{LSQ}^{\mathrm{ODEs}} RRL_1^n = \left(J^1(\mathbb{R}, M), \ \mathcal{LSQ} \right).$$

Definition 179. Any geometrical object on $J^1(\mathbb{R}, \mathbb{R}^n)$, which is produced by the jet least squares Lagrangian function $\mathcal{LSQ}^{\mathrm{ODEs}}$, via its attached second-order Euler-Lagrange equations, is called a *geometrical object produced by the jet first-order ODEs system (10.5)*.

By simple computations, the geometrical results of Theorems 176 and 177 lead us to the following

Theorem 180. (i) *The* **canonical nonlinear connection**

$$\Gamma_{\mathcal{S}_{\text{ODEs}}(\Delta)} = \left(M^{(i)}_{(1)1}, N^{(i)}_{(1)j} \right)$$

of the space $\mathcal{LSQ}^{\text{ODEs}} RRL^n_1$ *is given by the components*

$$M^{(i)}_{(1)1} = 0 \quad and \quad N^{(i)}_{(1)j} = -\frac{1}{2} \left[\frac{\partial X^{(i)}_{(1)}}{\partial x^j} - \frac{\partial X^{(j)}_{(1)}}{\partial x^i} \right].$$

(ii) *The* **canonical generalized Cartan connection** $C\Gamma_{\mathcal{S}_{\text{ODEs}}(\Delta)}$ *of the relativistic time-dependent Lagrange space* $\mathcal{LSQ}^{\text{ODEs}} RRL^n_1$ *has all its adapted components equal to zero.*

(iii) *The* **torsion** *d-tensor* **T** *of the canonical generalized Cartan connection* $C\Gamma_{\mathcal{S}_{\text{ODEs}}(\Delta)}$ *of the space* $\mathcal{LSQ}^{\text{ODEs}} RRL^n_1$ *is determined by* **two** *adapted local d-tensors:*

$$R^{(i)}_{(1)1j} = \frac{1}{2} \left[\frac{\partial^2 X^{(i)}_{(1)}}{\partial x^j \partial t} - \frac{\partial^2 X^{(j)}_{(1)}}{\partial x^i \partial t} \right] \quad and \quad R^{(i)}_{(1)jk} = -\frac{1}{2} \left[\frac{\partial^2 X^{(i)}_{(1)}}{\partial x^j \partial x^k} - \frac{\partial^2 X^{(j)}_{(1)}}{\partial x^i \partial x^k} \right].$$

(iv) *All adapted components of the* **curvature** *d-tensor* **R** *of the canonical generalized Cartan connection* $C\Gamma_{\mathcal{S}_{\text{ODEs}}(\Delta)}$ *of the space* $\mathcal{LSQ}^{\text{ODEs}} RRL^n_1$ *vanish.*

(v) *The* **electromagnetic** *2-form* $\mathbb{F}$ *of the space* $\mathcal{LSQ}^{\text{ODEs}} RRL^n_1$ *is expressed by*

$$\mathbb{F} = F^{(1)}_{(i)j} \delta y^i_1 \wedge dx^j,$$

where

$$\delta y^i_1 = dy^i_1 - \frac{1}{2} \left[\frac{\partial X^{(i)}_{(1)}}{\partial x^j} - \frac{\partial X^{(j)}_{(1)}}{\partial x^i} \right] dx^j \quad and \quad F^{(1)}_{(i)j} = \frac{1}{2} \left[\frac{\partial X^{(i)}_{(1)}}{\partial x^j} - \frac{\partial X^{(j)}_{(1)}}{\partial x^i} \right].$$

(vi) *The electromagnetic components* $F^{(1)}_{(i)j}$ *of the relativistic time-dependent Lagrange space* $\mathcal{LSQ}^{\text{ODEs}} RRL^n_1$ *are governed by the following* **geometrical Maxwell equations:**

$$\begin{cases} F^{(1)}_{(i)j//1} = \frac{1}{4} \mathcal{A}_{\{i,j\}} \left\{ \left[\frac{\partial^2 X^{(i)}_{(1)}}{\partial x^j \partial t} - \frac{\partial^2 X^{(j)}_{(1)}}{\partial x^i \partial t} \right] \right\} = \frac{1}{2} \left[\frac{\partial^2 X^{(i)}_{(1)}}{\partial x^j \partial t} - \frac{\partial^2 X^{(j)}_{(1)}}{\partial x^i \partial t} \right], \\ \\ \sum_{\{i,j,k\}} F^{(1)}_{(i)j\|k} = 0, \end{cases}$$

where $\mathcal{A}_{\{i,j\}}$ *represents an alternate sum,* $\sum_{\{i,j,k\}}$ *means a cyclic sum, and we have*

$$F_{(i)j//1}^{(1)} = \frac{\partial F_{(i)j}^{(1)}}{\partial t}, \quad F_{(i)j||k}^{(1)} = \frac{\partial F_{(i)j}^{(1)}}{\partial x^k}.$$

In our geometrical and physical approach, we point out that the distinguished electromagnetic 2-forms $\mathbb{F}$ produced by the ODEs systems and pairs of Euclidian metrics may be regarded as $o(n)$-valued 1-forms on the time manifold $\mathbb{R}$, setting

$$\mathbb{F} = \mathbf{F}_{(1)}dt \in \Gamma(\Lambda^1(T^*\mathbb{R}) \otimes o(n)),$$

where

$$\mathbf{F}_{(1)} = \begin{pmatrix} 0 & F_{(1)2}^{(1)} & F_{(1)3}^{(1)} & \cdots & \cdots & F_{(1)n}^{(1)} \\ -F_{(1)2}^{(1)} & 0 & F_{(2)3}^{(1)} & \cdots & \cdots & F_{(2)n}^{(1)} \\ -F_{(1)3}^{(1)} & -F_{(2)3}^{(1)} & 0 & \cdots & \cdots & F_{(3)n}^{(1)} \\ \cdots & \cdots & \cdots & \cdots & \cdots & \cdots \\ \cdots & \cdots & \cdots & \cdots & 0 & F_{(n-1)n}^{(1)} \\ -F_{(1)n}^{(1)} & -F_{(2)n}^{(1)} & -F_{(3)n}^{(1)} & \cdots & -F_{(n-1)n}^{(1)} & 0 \end{pmatrix} \in o(n).$$

Note that $o(n)$ is the set of skew-matrices as the Lie algebra $L(O(n))$ of the subgroup of orthogonal matrices $O(n) \subset GL_n(\mathbb{R})$. As a conclusion, we can introduce the following geometrical and abstract physical concept:

Definition 181. The Lagrangian function of Yang-Mills type, which is given by the formula

$$\mathcal{E}\mathcal{Y}\mathcal{M}_\Delta^{\mathrm{ODEs}}(\mathbb{F})(t,x) = \sum_{i=1}^{n-1}\sum_{j=i+1}^{n}\left[F_{(i)j}^{(1)}\right]^2 = \frac{1}{4}\sum_{i=1}^{n-1}\sum_{j=i+1}^{n}\left[\frac{\partial X_{(1)}^{(i)}}{\partial x^j} - \frac{\partial X_{(1)}^{(j)}}{\partial x^i}\right]^2,$$

is called the *jet geometric Yang-Mills energy* produced by the ODEs system (10.5).

Remark 182. If we use the matrix notations

- $J\left(X_{(1)}\right) = \left(\dfrac{\partial X_{(1)}^{(i)}}{\partial x^j}\right)_{i,j=\overline{1,n}}$ - the *Jacobian matrix,*

- $N_{(1)} = \left(N_{(1)j}^{(i)}\right)_{i,j=\overline{1,n}}$ - the *nonlinear connection matrix,*

- $R_{(1)1} = \left(R_{(1)1j}^{(i)}\right)_{i,j=\overline{1,n}}$, - the *temporal torsion matrix,*

- $R_{(1)k} = \left(R^{(i)}_{(1)jk} \right)_{i,j=\overline{1,n}}, \forall\, k = \overline{1,n}$, - the *spatial torsion matrices*,

- $\mathbf{F}_{(1)} := F^{(1)} = \left(F^{(1)}_{(i)j} \right)_{i,j=\overline{1,n}}$ - the *electromagnetic matrix*,

then the following matrix geometrical relations attached to the jet first-order ODEs system (10.5) hold true:

1. $N_{(1)} = -\dfrac{1}{2} \left[J\left(X_{(1)} \right) - {}^{T}J\left(X_{(1)} \right) \right]$;

2. $R_{(1)1} = -\dfrac{\partial}{\partial t} \left[N_{(1)} \right]$;

3. $R_{(1)k} = \dfrac{\partial}{\partial x^k} \left[N_{(1)} \right], \forall\, k = \overline{1,n}$;

4. $F^{(1)} = -N_{(1)}$;

5. $\mathcal{EYM}^{\mathrm{ODEs}}_{\Delta}(t,x) = \dfrac{1}{2} \cdot \mathrm{Trace} \left[F^{(1)} \cdot {}^{T}F^{(1)} \right]$,

   that is the jet geometric Yang-Mills energy [produced by the ODEs system (10.5)] coincides with the square of the norm of the skew-symmetric electromagnetic matrix $F^{(1)}$ in the Lie algebra $o(n) = L(O(n))$.

In many problems from Applied Sciences we meet d-tensors $X^{(i)}_{(1)}$ which are not dependent on the time coordinate $t \in \mathbb{R}$. In other words, in many applicative problems we are working with ODEs systems of order one given by d-tensors X on $J^1(\mathbb{R}, \mathbb{R}^n)$ whose components have the particular form

$$X^{(i)}_{(1)} = X^{(i)}_{(1)}(x^k).$$

In these time-independent situations, it is obvious that many geometrical objects studied by us cancel. In fact, for time-independent d-tensors $X^{(i)}_{(1)} = X^{(i)}_{(1)}(x^k)$, we have the ODEs system of first-order (*jet dynamical system*)

$$\frac{dx^i}{dt} = X^{(i)}_{(1)}(x^k(t)), \quad \forall\, i = \overline{1,n}. \tag{10.14}$$

Using Theorem 180, we can assert that the Riemann-Lagrange geometrical objects produced by the ODEs system (10.14) and the pair of Euclidian metrics Δ are given in the following result:

Corollary 183. (i) *The* **canonical nonlinear connection**

$$\Gamma = \left(M^{(i)}_{(1)1}, N^{(i)}_{(1)j} \right),$$

produced by the ODEs system (10.14), *is given by the components*

$$M^{(i)}_{(1)1} = 0 \quad and \quad N^{(i)}_{(1)j} = -\frac{1}{2}\left[\frac{\partial X^{(i)}_{(1)}}{\partial x^j} - \frac{\partial X^{(j)}_{(1)}}{\partial x^i}\right].$$

(ii) *All adapted components of the* **canonical generalized Cartan connection** $C\Gamma$, **produced by the ODEs system (10.14)**, *are zero.*

(iii) *The* **torsion** *d-tensor* **T** *of the canonical generalized Cartan connection* $C\Gamma$, **produced by the ODEs system (10.14)**, *is determined only by the adapted components*

$$R^{(i)}_{(1)jk} = -\frac{1}{2}\left[\frac{\partial^2 X^{(i)}_{(1)}}{\partial x^j \partial x^k} - \frac{\partial^2 X^{(j)}_{(1)}}{\partial x^i \partial x^k}\right].$$

(iv) *All adapted components of the* **curvature** *d-tensor* **R, produced by the ODEs system (10.14)**, *are zero.*

(v) *The* **electromagnetic 2-form** $\mathbb{F}$, **produced by the ODEs system (10.14)**, *is expressed by*

$$\mathbb{F} = F^{(1)}_{(i)j}\delta y^i_1 \wedge dx^j,$$

where

$$\delta y^i_1 = dy^i_1 - \frac{1}{2}\left[\frac{\partial X^{(i)}_{(1)}}{\partial x^j} - \frac{\partial X^{(j)}_{(1)}}{\partial x^i}\right]dx^j \quad and \quad F^{(1)}_{(i)j} = \frac{1}{2}\left[\frac{\partial X^{(i)}_{(1)}}{\partial x^j} - \frac{\partial X^{(j)}_{(1)}}{\partial x^i}\right].$$

(vi) *The components* $F^{(1)}_{(i)j}$ *are governed by the following simpler* **geometrical Maxwell equations***:*

$$\sum_{\{i,j,k\}} F^{(1)}_{(i)j\|k} = 0.$$

(vii) *The* **jet geometric Yang-Mills energy, produced by the jet first-order ODEs system (10.14)**, *has the expression*

$$\mathcal{EYM}^{\mathrm{ODEs}}_{\Delta}(x) = \sum_{i=1}^{n-1}\sum_{j=i+1}^{n}\left[F^{(1)}_{(i)j}\right]^2 = \frac{1}{4}\sum_{i=1}^{n-1}\sum_{j=i+1}^{n}\left[\frac{\partial X^{(i)}_{(1)}}{\partial x^j} - \frac{\partial X^{(j)}_{(1)}}{\partial x^i}\right]^2.$$

Remark 184. Using the matrix notations from Remark 182, the following matrix geometrical objects characterize the jet first-order ODEs system (10.14):

1. $N_{(1)} = -\dfrac{1}{2}\left[J\left(X_{(1)}\right) - {}^T J\left(X_{(1)}\right)\right]$;

2. $R_{(1)k} = \dfrac{\partial}{\partial x^k}\left[N_{(1)}\right], \forall\, k = \overline{1,n}$;

3. $F^{(1)} = -N_{(1)}$;

4. $\mathcal{EYM}^{\mathrm{ODEs}}_{\Delta}(x) = \dfrac{1}{2} \cdot \mathrm{Trace}\left[F^{(1)} \cdot {}^T F^{(1)}\right]$.

CHAPTER 11

JET SINGLE-TIME LAGRANGE GEOMETRY APPLIED TO THE LORENZ ATMOSPHERIC ODEs SYSTEM

In this Chapter, using the general formulas from Corollary 183 and Remark 184, we apply the Riemann-Lagrange geometrical results from Chapter 10 to the Lorenz five-components atmospheric ODEs system introduced by Lorenz [49] and studied, via the Melnikov function method for Hamiltonian systems on Lie groups, by Birtea, Puta, Rațiu, and Tudoran [21].

11.1 JET RIEMANN-LAGRANGE GEOMETRY PRODUCED BY THE LORENZ SIMPLIFIED MODEL OF ROSSBY GRAVITY WAVE INTERACTION

The first model equations for the atmosphere are that so called *primitive equations*. It seems that this model produces wave-like motions on different time scales:

- on the one hand, this model produces the slow motions which have a period of order of days (these slow-waves are called *Rossby waves*);

- on the other hand, this model produces fast motions which have a period of hours (these fast-waves are called *gravity waves*).

Jet Single-Time Lagrange Geometry and Its Applications
1-st Edition. By Vladimir Balan and Mircea Neagu.
© 2011 John Wiley & Sons, Inc. Published 2011 John Wiley & Sons, Inc.

The question of how to balance these two time scales led Lorenz [49] to consider a simplified version of the primitive equations model, which is given by the following nonlinear system of five differential equations [21]:

$$
\begin{cases}
\dfrac{dx^1}{dt} = -x^2 x^3 + \varepsilon x^2 x^5, \\[2mm]
\dfrac{dx^2}{dt} = x^1 x^3 - \varepsilon x^1 x^5, \\[2mm]
\dfrac{dx^3}{dt} = -x^1 x^2, \\[2mm]
\dfrac{dx^4}{dt} = -x^5, \\[2mm]
\dfrac{dx^5}{dt} = x^4 + \varepsilon x^1 x^2,
\end{cases}
\tag{11.1}
$$

where the variables x^4, x^5 represent the fast gravity wave oscillations and the variables x^1, x^2, x^3 are the slow Rossby wave oscillations, with a parameter ε which is related to the physical Rossby number.

Remark 185. It is obvious that, from a physical point of view, the Lorenz atmospheric ODEs system (11.1) couples the Rossby waves with the gravity waves.

Naturally, the Lorenz atmospheric ODEs system (11.1) can be regarded as a nonlinear ODEs system of order one on the 1-jet space $J^1(\mathbb{R}, \mathbb{R}^5)$, which is produced by the d-tensor field $X = \left(X_{(1)}^{(i)}(x) \right)$, where $i = \overline{1,5}$ and $x = (x^1, x^2, x^3, x^4, x^5)$, having the local components

$$
\begin{aligned}
X_{(1)}^{(1)}(x) &= -x^2 x^3 + \varepsilon x^2 x^5, \\
X_{(1)}^{(2)}(x) &= x^1 x^3 - \varepsilon x^1 x^5, \\
X_{(1)}^{(3)}(x) &= -x^1 x^2, \\
X_{(1)}^{(4)}(x) &= -x^5, \\
X_{(1)}^{(5)}(x) &= x^4 + \varepsilon x^1 x^2.
\end{aligned}
\tag{11.2}
$$

Consequently, via the Corollary 183, we assert that the Riemann-Lagrange geometrical behavior on the 1-jet space $J^1(\mathbb{R}, \mathbb{R}^5)$ of the Lorenz atmospheric ODEs system (11.1) is described in the following result (see also Remark 184):

Theorem 186. (i) *The* **canonical nonlinear connection on** $J^1(\mathbb{R}, \mathbb{R}^5)$, **produced by the Lorenz atmospheric ODEs system (11.1)**, *has the local components*

$$
\hat{\Gamma} = \left(0, \hat{N}_{(1)j}^{(i)} \right),
$$

where $\hat{N}_{(1)j}^{(i)}$ are the entries of the matrix

$$\hat{N}_{(1)} = \left(\hat{N}_{(1)j}^{(i)}\right)_{i,j=\overline{1,5}} = \begin{pmatrix} 0 & x^3 - \varepsilon x^5 & 0 & 0 & 0 \\ -x^3 + \varepsilon x^5 & 0 & -x^1 & 0 & \varepsilon x^1 \\ 0 & x^1 & 0 & 0 & 0 \\ 0 & 0 & 0 & 0 & 1 \\ 0 & -\varepsilon x^1 & 0 & -1 & 0 \end{pmatrix}.$$

(ii) *All adapted components of the* **canonical generalized Cartan connection** $C\hat{\Gamma}$, **produced by the Lorenz atmospheric ODEs system (11.1)**, *are zero.*

(iii) *All adapted components of the* **torsion** *d-tensor* $\hat{T}$ *of the canonical generalized Cartan connection* $C\hat{\Gamma}$, **produced by the Lorenz atmospheric ODEs system (11.1)**, *are zero, except*

$$\hat{R}_{(1)21}^{(3)} = -\hat{R}_{(1)31}^{(2)} = 1, \qquad \hat{R}_{(1)21}^{(5)} = -\hat{R}_{(1)51}^{(2)} = -\varepsilon,$$

$$\hat{R}_{(1)13}^{(2)} = -\hat{R}_{(1)23}^{(1)} = -1, \qquad \hat{R}_{(1)15}^{(2)} = -\hat{R}_{(1)25}^{(1)} = \varepsilon.$$

(iv) *All adapted components of the* **curvature** *d-tensor* $\hat{R}$ *of the canonical generalized Cartan connection* $C\hat{\Gamma}$, **produced by the Lorenz atmospheric ODEs system (11.1)**, *cancel.*

(v) *The* **geometric electromagnetic distinguished 2-form**, **produced by the Lorenz atmospheric ODEs system (11.1)**, *has the expression*

$$\hat{\mathbb{F}} = \hat{F}_{(i)j}^{(1)} \delta y_1^i \wedge dx^j,$$

where

$$\delta y_1^i = dy_1^i + \hat{N}_{(1)k}^{(i)} dx^k, \ \forall \, i = \overline{1,5},$$

and the adapted components $\hat{F}_{(i)j}^{(1)}$ *are the entries of the matrix*

$$\hat{F}^{(1)} = \left(\hat{F}_{(i)j}^{(1)}\right)_{i,j=\overline{1,5}} = -\hat{N}_{(1)}.$$

(vi) *The* **jet geometric Yang-Mills energy**, **produced by the Lorenz atmospheric ODEs system (11.1)**, *is given by the formula*

$$\mathcal{EYM}^{\text{Lorenz}}(x) = \left(\varepsilon x^5 - x^3\right)^2 + \left(x^1\right)^2 + \left(\varepsilon x^1\right)^2 + 1.$$

Proof: The Lorenz atmospheric ODEs system (11.1) is a particular case of the jet first-order ODEs system (10.14) for $n = 5$ and $X = \left(X_{(1)}^{(i)}(x)\right)_{i=\overline{1,5}}$ given by relations (11.2). In conclusion, applying Corollary 183 (together with Remark 184) and using the Jacobian matrix

$$J\left(X_{(1)}\right) = \begin{pmatrix} 0 & -x^3 + \varepsilon x^5 & -x^2 & 0 & \varepsilon x^2 \\ x^3 - \varepsilon x^5 & 0 & x^1 & 0 & -\varepsilon x^1 \\ -x^2 & -x^1 & 0 & 0 & 0 \\ 0 & 0 & 0 & 0 & -1 \\ \varepsilon x^2 & \varepsilon x^1 & 0 & 1 & 0 \end{pmatrix},$$

we obtain what we were looking for.

Remark 187. Let us remark that, although the jet geometric Yang-Mills energy $\mathcal{EYM}^{\text{Lorenz}}$, produced by the Lorenz atmospheric ODEs system (11.1), depends only by the coordinates x^1, x^3, and x^5, it still couples the slow Rossby wave oscillations with the fast gravity wave oscillations. However, the coordinates x^2 and x^4 are missing in the expression of $\mathcal{EYM}^{\text{Lorenz}}$. An open question is whether there exists a physical interpretation of this fact.

11.2 YANG-MILLS ENERGETIC HYPERSURFACES OF CONSTANT LEVEL PRODUCED BY THE LORENZ ATMOSPHERIC ODEs SYSTEM

In the preceding Riemann-Lagrange geometrical theory on the 1-jet space $J^1(\mathbb{R}, \mathbb{R}^5)$ the Lorenz atmospheric ODEs system (11.1) "produces" a jet Yang-Mills energy given by the formula

$$\mathcal{EYM}^{\text{Lorenz}}(x) = \left(1 + \varepsilon^2\right)\left(x^1\right)^2 + \left(x^3\right)^2 + \varepsilon^2(x^5)^2 - 2\varepsilon x^3 x^5 + 1,$$

where $x = (x^1, x^2, x^3, x^4, x^5)$. In what follows, we study the *jet geometric Yang-Mills energetic hypersurfaces of constant level*, produced by the Lorenz atmospheric ODEs system (11.1), which are defined by the implicit equations

$$\Sigma_C^{\text{Lorenz}} : \left(\varepsilon x^5 - x^3\right)^2 + \left(1 + \varepsilon^2\right)\left(x^1\right)^2 = C - 1,$$

where C is a constant real number.

Because Σ_C^{Lorenz} is a *quadric* in the system of axes $Ox^1x^3x^5$ for every $C \in \mathbb{R}$, then, using the reduction to the canonical form of a quadric, we find the following geometrical results:

1. If $C < 1$, then we have $\Sigma_{C<1}^{\text{Lorenz}} = \emptyset$.

2. If $C = 1$, then we have

$$\Sigma_{C=1}^{\text{Lorenz}} : \begin{cases} x^1 = 0, \\ x^3 - \varepsilon x^5 = 0, \end{cases}$$

that is, $\Sigma_{C=1}^{\text{Lorenz}}$ is a *straight line* in the system of axes $Ox^1x^3x^5$.

3. If $C > 1$, then we have

$$\Sigma_{C>1}^{\text{Lorenz}} : \left(1 + \varepsilon^2\right)\left(x^1\right)^2 + \left(x^3\right)^2 + \varepsilon^2(x^5)^2 - 2\varepsilon x^3 x^5 - C + 1 = 0,$$

that is, $\Sigma_{C>1}^{\text{Lorenz}}$ is a degenerate non-empty quadric in the system of axes $Ox^1x^3x^5$, whose canonical form is

$$\Sigma_{C>1}^{\text{Lorenz}} : \left(X^3\right)^2 + \left(X^5\right)^2 = \frac{C-1}{1+\varepsilon^2}.$$

Note that the rotation of the system of axes $Ox^1x^3x^5$ into the system of axes $OX^1X^3X^5$ is given by the matrix relation

$$\begin{pmatrix} x^1 \\ x^3 \\ x^5 \end{pmatrix} = \frac{1}{\sqrt{1+\varepsilon^2}} \begin{pmatrix} 0 & \sqrt{1+\varepsilon^2} & 0 \\ \varepsilon & 0 & 1 \\ 1 & 0 & -\varepsilon \end{pmatrix} \begin{pmatrix} X^1 \\ X^3 \\ X^5 \end{pmatrix}.$$

In conclusion, the degenerate non-empty quadric $\Sigma_{C>1}^{\text{Lorenz}}$ is in the system of axes $Ox^1x^3x^5$ a *slant circular cylinder* of radius

$$R = \sqrt{\frac{C-1}{1+\varepsilon^2}},$$

having as an axis of symmetry the straight line $\Sigma_{C=1}^{\text{Lorenz}}$.

Open problem. Find physical interpretations for our preceding geometrical results regarding the study of the Lorenz atmospheric ODEs system (11.1).

CHAPTER 12

JET SINGLE-TIME LAGRANGE GEOMETRY APPLIED TO EVOLUTION ODes SYSTEMS FROM ECONOMY

Using again the general results from Chapter 10, in the next Sections, we apply our jet Riemann-Lagrange geometrical results to certain evolution equations that govern certain phenomena in Economy, extending in this way the geometrical studies initiated by Udrişte, Ferrara, and Opriş in the book [95]. In fact, we will study the jet Riemann-Lagrange geometry produced by the following two economic mathematical models:

- the Kaldor nonlinear model of the business cycle (cf. Gandolfo [35]);

- the Tobin-Benhabib-Miyao economic model regarding the role of currency on economic growth (cf. Tobin [93] and Benhabib-Miyao [22]).

12.1 JET RIEMANN-LAGRANGE GEOMETRY FOR KALDOR NONLINEAR CYCLICAL MODEL IN BUSINESS

The *national revenue* $Y(t)$ and the *capital stock* $K(t)$, where $t \in [a, b]$, are the state variables of the Kaldor nonlinear model of the business cycle. The kinetic Kaldor model of a commercial cycle belongs to the category of business cycles described by the *Kaldor flow* (for more details, see [35] and [95])

$$
\begin{cases}
\dfrac{dY}{dt} = s\left[I(Y,K) - S(Y,K)\right], \\[2mm]
\dfrac{dK}{dt} = I(Y,K) - qK,
\end{cases}
\tag{12.1}
$$

where

- $I = I(Y, K)$ is a given differentiable *investment function*, which verifies some *economic-mathematical Kaldor conditions*;

- $S = S(Y, K)$ is a given differentiable *saving function*, which verifies some *economic-mathematical Kaldor conditions*;

- $s > 0$ is an adjustement constant parameter which measures the reaction of the model with respect to the difference between the investment function and the saving function;

- $q \in (0, 1)$ is a constant representing the depreciation coefficient of capital.

Remark 188. Details of the *Kaldor economic-mathematical conditions* imposed on the given functions I and S are found in [35] and [95]. We underline that, from the point of view of our jet Riemann-Lagrange geometry produced by the Kaldor evolution equations (geometry that we will describe in the sequel) the economic-mathematical hypotheses of Kaldor can be neglected because all our necessary geometrical informations are concentrated in the Kaldor flow (12.1).

The Riemann-Lagrange geometrical behavior on the 1-jet space $J^1([a, b], \mathbb{R}^2)$ of the Kaldor economic evolution model is described in the following result (cf. Corollary 183 and Remark 184):

Theorem 189. (i) *The* **canonical nonlinear connection on** $J^1([a, b], \mathbb{R}^2)$, **produced by the Kaldor flow (12.1)**, *has the local components*

$$
\hat{\Gamma} = \left(0, \hat{N}_{(1)j}^{(i)}\right),
$$

where, if I_Y, I_K, *and* S_K *are the partial derivatives of the functions* I *and* S, *then* $\hat{N}_{(1)j}^{(i)}$ *are the entries of the matrix*

$$
\hat{N}_{(1)} =
\begin{pmatrix}
0 & \dfrac{1}{2}[I_Y - s(I_K - S_K)] \\[3mm]
-\dfrac{1}{2}[I_Y - s(I_K - S_K)] & 0
\end{pmatrix}.
$$

(ii) *All adapted components of the* **canonical generalized Cartan connection** $C\hat{\Gamma}$, **produced by the Kaldor flow (12.1)**, *are zero.*

(iii) *All adapted components of the* **torsion** *d-tensor* $\hat{\mathbf{T}}$ *of the canonical generalized Cartan connection* $C\hat{\Gamma}$, **produced by the Kaldor flow (12.1)**, *are zero, except*

$$\hat{R}^{(1)}_{(1)21} = -\hat{R}^{(2)}_{(1)11} = \frac{1}{2}[I_{YY} - s(I_{YK} - S_{YK})],$$

$$\hat{R}^{(1)}_{(1)22} = -\hat{R}^{(2)}_{(1)12} = \frac{1}{2}[I_{YK} - s(I_{KK} - S_{KK})],$$

where I_{YY}, I_{YK}, I_{KK}, S_{YK}, *and* S_{KK} *are the second partial derivatives of the given functions* I *and* S.

(iv) *All adapted components of the* **curvature** *d-tensor* $\hat{\mathbf{R}}$ *of the canonical generalized Cartan connection* $C\hat{\Gamma}$, **produced by the Kaldor flow (12.1)**, *cancel.*

(v) *The geometric* **electromagnetic distinguished 2-form**, **produced by the Kaldor flow (12.1)**, *has the expression*

$$\hat{\mathbb{F}} = \hat{F}^{(1)}_{(i)j}\delta y^i_1 \wedge dx^j,$$

where

$$\delta y^i_1 = dy^i_1 + \hat{N}^{(i)}_{(1)k}dx^k, \ \forall \ i = \overline{1,2},$$

and the adapted components $\hat{F}^{(1)}_{(i)j}$ *are the entries of the matrix*

$$\hat{F}^{(1)} = -\hat{N}_{(1)} = \begin{pmatrix} 0 & -\dfrac{1}{2}[I_Y - s(I_K - S_K)] \\ \dfrac{1}{2}[I_Y - s(I_K - S_K)] & 0 \end{pmatrix}.$$

(vi) *The* **economic geometric Yang-Mills energy**, **produced by the Kaldor flow (12.1)**, *is given by the formula*

$$\mathcal{EYM}^{\text{Kaldor}}(Y,K) = \frac{1}{4}[I_Y - s(I_K - S_K)]^2.$$

Proof: Let us regard the Kaldor flow (12.1) as a particular case of the jet first-order ODEs system (10.14) on the 1-jet space $J^1([a,b],\mathbb{R}^2)$, taking

$$n = 2, \quad x^1 = Y, \quad x^2 = K,$$

and putting

$$X^{(1)}_{(1)}(x^1,x^2) = s\left[I(x^1,x^2) - S(x^1,x^2)\right], \qquad X^{(2)}_{(1)}(x^1,x^2) = I(x^1,x^2) - qx^2.$$

Now, taking into account that we have the Jacobian matrix

$$J\left(X_{(1)}\right) = \begin{pmatrix} s\left[I_{x^1} - S_{x^1}\right] & s\left[I_{x^2} - S_{x^2}\right] \\ I_{x^1} & I_{x^2} - q \end{pmatrix}$$

$$= \begin{pmatrix} s\left[I_Y - S_Y\right] & s\left[I_K - S_K\right] \\ I_Y & I_K - q \end{pmatrix},$$

and using Corollary 183 and Remark 184, we obtain what we were looking for.

Open problem. The *geometric Yang-Mills economic energetic curves of constant level produced by the Kaldor flow (12.1)*, which are different by the empty set, are the curves in the plane YOK having the implicit equations

$$\Gamma_C : [I_Y - s(I_K - S_K)]^2 = 4C,$$

where $C \geq 0$. Determine the relevance of the shapes of the plane curves Γ_C for real economic processes.

12.2   JET RIEMANN-LAGRANGE GEOMETRY FOR TOBIN-BENHABIB-MIYAO ECONOMIC EVOLUTION MODEL

The Tobin mathematical model [93] regarding the role of money on economic growth was extended by Benhabib and Miyao [22] by incorporating the role of some expectation constant parameters. Thus, the Tobin-Benhabib-Miyao (TBM) economic model relies on the variables $k(t)$ = *the capital labor ratio*, $m(t)$ = *the money stock per head*, $q(t)$ = *the expected rate of inflation*, whose evolution in time is given by the *TBM flow* [95]

$$\begin{cases} \dfrac{dk}{dt} = sf(k(t)) - (1-s)[\theta - q(t)]m(t) - nk(t) \\[2mm] \dfrac{dm}{dt} = m(t)\left\{\theta - n - q(t) - \varepsilon[m(t) - l(k(t), q(t))]\right\} & (12.2) \\[2mm] \dfrac{dq}{dt} = \mu\varepsilon[m(t) - l(k(t), q(t))], \end{cases}$$

where the $f(k)$ and $l(k, q)$ are some given differentiable real functions and s, θ, n, μ, ε are expectation parameters: s = *saving ratio*, θ = *rate of money expansion*, n = *population growth rate*, μ = *speed of adjustement of expectations*, ε = *speed of adjustement of price level*.

Remark 190. From the point of view of economists, the *actual rate of inflation* in the TBM economic model is given by the formula [95]

$$\overline{p}(t) = \varepsilon[m(t) - l(k(t), q(t))] + q(t).$$

In what follows, we apply our jet Riemann-Lagrange geometrical results to the TBM flow (12.2). In this context, we obtain (cf. Corollary 183 and Remark 184) the following:

Theorem 191. (i) *The* **canonical nonlinear connection** *on* $J^1([a, b], \mathbb{R}^3)$, **produced by the TBM flow (12.2)**, *has the local components*

$$\breve{\Gamma} = \left(0, \breve{N}^{(i)}_{(1)j}\right),$$

where, if l_k and l_q are the partial derivatives of the function l, then $\check{N}^{(i)}_{(1)j}$ are the entries of the matrix

$$
\check{N}_{(1)} = -\frac{1}{2}
\begin{pmatrix}
0 & \begin{matrix} -(1-s)(\theta-q)- \\ -\varepsilon m l_k \end{matrix} & (1-s)m + \mu\varepsilon l_k \\
\begin{matrix} (1-s)(\theta-q)+ \\ +\varepsilon m l_k \end{matrix} & 0 & -m + \varepsilon m l_q - \mu\varepsilon \\
-(1-s)m - \mu\varepsilon l_k & m - \varepsilon m l_q + \mu\varepsilon & 0
\end{pmatrix}.
$$

(ii) *All adapted components of the* **canonical generalized Cartan connection** $C\check{\Gamma}$, **produced by the TBM flow (12.2)**, *are zero*.

(iii) *The effective adapted components of the* **torsion** *d-tensor* $\check{T}$ *of the canonical generalized Cartan connection* $C\check{\Gamma}$, **produced by the TBM flow (12.2)**, *are the entries of the matrices*

$$
\check{R}_{(1)1} = -\frac{1}{2}
\begin{pmatrix}
0 & -\varepsilon m l_{kk} & \mu\varepsilon l_{kk} \\
\varepsilon m l_{kk} & 0 & \varepsilon m l_{kq} \\
-\mu\varepsilon l_{kk} & -\varepsilon m l_{kq} & 0
\end{pmatrix},
$$

$$
\check{R}_{(1)2} = -\frac{1}{2}
\begin{pmatrix}
0 & -\varepsilon l_k & 1-s \\
\varepsilon l_k & 0 & -1+\varepsilon l_q \\
-1+s & 1-\varepsilon l_q & 0
\end{pmatrix},
$$

and

$$
\check{R}_{(1)3} = -\frac{1}{2}
\begin{pmatrix}
0 & 1-s-\varepsilon m l_{kq} & \mu\varepsilon l_{kq} \\
-1+s+\varepsilon m l_{kq} & 0 & \varepsilon m l_{qq} \\
-\mu\varepsilon l_{kq} & -\varepsilon m l_{qq} & 0
\end{pmatrix},
$$

where l_{kk}, l_{kq}, and l_{qq} are the second partial derivatives of the function l and

$$
\check{R}_{(1)k} = \left(\check{R}^{(i)}_{(1)jk} \right)_{i,j=\overline{1,3}}, \ \forall \, k = \overline{1,3}.
$$

(iv) *All adapted components of the* **curvature** *d-tensor* $\check{R}$ *of the canonical generalized Cartan connection* $C\check{\Gamma}$, **produced by the TBM flow (12.2)**, *cancel*.

(v) *The geometric* **electromagnetic distinguished** 2-**form, produced by the TBM flow (12.2)**, *has the expression*

$$
\mathbb{\check{F}} = \check{F}^{(1)}_{(i)j} \delta y^i_1 \wedge dx^j,
$$

where

$$
\delta y^i_1 = dy^i_1 + \check{N}^{(i)}_{(1)k} dx^k, \ \forall \, i = \overline{1,3},
$$

and the adapted components $\check{F}^{(1)}_{(i)j}$ are the entries of the matrix

$$
\check{F}^{(1)} = -\check{N}_{(1)}.
$$

(vi) *The* **economic geometric Yang-Mills energy, produced by the TBM flow** **(12.2)**, *is given by the formula*

$$\mathcal{E}\mathcal{Y}\mathcal{M}^{\text{TBM}}(k, m, q) = \frac{1}{4}\left\{[(1-s)(\theta - q) + \varepsilon m l_k]^2 + [(1-s)m + \mu\varepsilon l_k]^2\right.$$
$$\left. + [m - \varepsilon m l_q + \mu\varepsilon]^2\right\}.$$

Proof: We regard the TBM flow (12.2) as a particular case of the jet first-order ODEs system (10.14) on the 1-jet space $J^1([a, b], \mathbb{R}^3)$, taking

$$n = 3, \ x^1 = k, \ x^2 = m, \ x^3 = q$$

and putting

$$X_{(1)}^{(1)}(x^1, x^2, x^3) = sf(x^1) - (1-s)[\theta - x^3]x^2 - nx^1,$$

$$X_{(1)}^{(2)}(x^1, x^2, x^3) = x^2\left\{\theta - n - x^3 - \varepsilon[x^2 - l(x^1, x^3)]\right\},$$

and

$$X_{(1)}^{(3)}(x^1, x^2, x^3) = \mu\varepsilon[x^2 - l(x^1, x^3)].$$

It follows that we have the Jacobian matrix

$$J\left(X_{(1)}\right) = \begin{pmatrix} sf'(x^1) - n & -(1-s)(\theta - x^3) & (1-s)x^2 \\ \varepsilon x^2 l_{x^1} & \begin{array}{c} -2\varepsilon x^2 + \theta - x^3 - n+ \\ +\varepsilon l(x^1, x^3) \end{array} & -x^2 + \varepsilon x^2 l_{x^3} \\ -\mu\varepsilon l_{x^1} & \mu\varepsilon & -\mu\varepsilon l_{x^3} \end{pmatrix}$$

$$= \begin{pmatrix} sf'(k) - n & -(1-s)(\theta - q) & (1-s)m \\ \varepsilon m l_k & \begin{array}{c} -2\varepsilon m + \theta - q - n+ \\ +\varepsilon l(k, q) \end{array} & -m + \varepsilon m l_q \\ -\mu\varepsilon l_k & \mu\varepsilon & -\mu\varepsilon l_q \end{pmatrix},$$

where f' is the derivative of the function f. In conclusion, using Corollary 183 and Remark 184, we find the required result.

Open problem.   The *Yang-Mills economic energetic surfaces of constant level produced by the TBM flow (12.2)*, which are different by the empty set, have in the system of axis $Okmq$ the implicit equations

$$\Sigma_C : [(1-s)(\theta - q) + \varepsilon m l_k]^2 + [(1-s)m + \mu\varepsilon l_k]^2 + [m - \varepsilon m l_q + \mu\varepsilon]^2 = 4C,$$

where $C \geq 0$. Find economic interpretations for the geometry of the Σ_C surfaces.

CHAPTER 13

SOME EVOLUTION EQUATIONS FROM THEORETICAL BIOLOGY AND THEIR SINGLE-TIME LAGRANGE GEOMETRIZATION ON 1-JET SPACES

Using the applicative results from Chapter 10 (particularly the results from Corollary 183 and Remark 184), the aim of this Chapter is to construct the Riemann-Lagrange differential geometry on 1-jet spaces (in the sense of nonlinear connections, generalized Cartan linear connections, d-torsions, d-curvatures, jet electromagnetic fields and jet electromagnetic Yang-Mills energies), starting from some given nonlinear evolution ODEs systems modeling biologic phenomena such as:

- the model of the cancer cell population (see [38] and [91]);

- the model of the infection by human immunodeficiency virus-type 1 (HIV-1) model (see [83]);

- the model of cytosolic calcium oscillations in living cells (see [42]);

- the model of calcium oscillations in living cells, through endoplasmic reticulum, mitochondria and cytosolic proteins (see [50]).

For more theoretical-biologic details, see the works of Nicola [74], [75], and [68].

Jet Single-Time Lagrange Geometry and Its Applications
1-st Edition. By Vladimir Balan and Mircea Neagu.
© 2011 John Wiley & Sons, Inc. Published 2011 John Wiley & Sons, Inc.

13.1   JET RIEMANN-LAGRANGE GEOMETRY FOR A CANCER CELL POPULATION MODEL IN BIOLOGY

The mathematical model of cancer cell population, which consists of a two-dimensional ODEs system having four parameters, was introduced in 2006 by Garner *et al.* (see [38]).

The cancer cell populations consist of a combination of proliferating, quiescent, and dead cells that determine tumor growth or cancer spread. Moreover, recent research in cancer progression and treatment indicates that many forms of cancer arise from one abnormal cell or a small subpopulation of abnormal cells. These cells, which support cancer growth and spread, are called *cancer stem cells* (CSCs). Targeting these CSCs is crucial because they display many of the same characteristics as healthy stem cells, and they have the capacity of initiating new tumors after long periods of remission. The understanding of the cancer mechanism could have a significant impact on cancer treatment approaches as it emphasizes the importance of targeting diverse cell subpopulations at a specific stage of development.

The non-dimensionalized model introduced by Garner *et al.* is based on a system of Solyanik *et al.* (see [91]), which starts from the following assumptions:

1. the cancer cell population consists of proliferating and quiescent (resting) cells;

2. the cells can lose their ability to divide under certain conditions and then transit from the proliferating to the resting state;

3. resting cells can either return to the proliferating state or die.

The dynamical system has two state variables, namely, P is *the number of proliferating cells* and Q is *the number of quiescent cells*, and their evolution in time is described by the following differential equations (*cancer cell population flow*):

$$\begin{cases} \dfrac{dP}{dt} = P - P(P + Q) + F(P,Q), \\[2mm] \dfrac{dQ}{dt} = -rQ + aP(P + Q) - F(P,Q), \end{cases} \tag{13.1}$$

$$F(P,Q) = \frac{hPQ}{1 + kP^2}, \quad r = \frac{d}{b}, \quad h = \frac{A}{ac}, \quad k = \frac{Bb^2}{c^2},$$

where

- a is a dimensionless constant that measures the relative nutrient uptake by resting and proliferating cells;

- b is the rate of cell division of the proliferating cells;

- c depends on the intensity of consumption by proliferating cells and gives the magnitude of the rate of cell transition from the proliferating stage to the resting stage in per cell per day;

- d is the rate of cell death of the resting cells (per day);

- A represents the initial rate of increase in the intensity of cell transition from the quiescent to proliferating state at small P;

- A/B represents the rate of decrease in the intensity of cell transition from the quiescent to proliferating state when P becomes larger.

The Riemann-Lagrange geometrical behavior on the 1-jet space $J^1(\mathbb{R}, \mathbb{R}^2)$ of the *cancer cell population flow* is described in the following result (cf. Corollary 183 and Remark 184):

Theorem 192. (i) *The* **canonical nonlinear connection** *on* $J^1(\mathbb{R}, \mathbb{R}^2)$, **produced by the cancer cell population flow (13.1)**, *has the local components*

$$\hat{\Gamma} = \left(0, \hat{N}^{(i)}_{(1)j}\right), \ i,j = \overline{1,2},$$

where, if

$$F_P = \frac{hQ\left(1 - kP^2\right)}{\left(1 + kP^2\right)^2} \ and \ F_Q = \frac{hP}{1 + kP^2}$$

are the first partial derivatives of the function F, then we have

$$\hat{N}^{(1)}_{(1)1} = \hat{N}^{(2)}_{(1)2} = 0,$$

$$\hat{N}^{(1)}_{(1)2} = -\hat{N}^{(2)}_{(1)1} = \frac{1}{2}\left[(2a + 1)P + aQ - (F_P + F_Q)\right]$$

$$= \frac{1}{2}\left[(2a + 1)P + aQ - \frac{hQ\left(1 - kP^2\right)}{\left(1 + kP^2\right)^2} - \frac{hP}{1 + kP^2}\right].$$

(ii) *All adapted components of the* **canonical generalized Cartan connection** $C\hat{\Gamma}$, **produced by the cancer cell population flow (13.1)**, *are zero.*

(iii) *All adapted components of the* **torsion d-tensor** $\hat{T}$ *of the canonical generalized Cartan connection* $C\hat{\Gamma}$, **produced by the cancer cell population flow (13.1)**, *are zero, except*

$$\hat{R}^{(1)}_{(1)21} = -\hat{R}^{(2)}_{(1)11} = a + \frac{1}{2}\left(1 - F_{PP} - F_{PQ}\right),$$

$$\hat{R}^{(1)}_{(1)22} = -\hat{R}^{(2)}_{(1)12} = \frac{1}{2}\left(a - F_{PQ} - F_{QQ}\right) = \frac{1}{2}\left(a - F_{PQ}\right),$$

where

$$F_{PP} = -\frac{2hkPQ\left(3 - kP^2\right)}{\left(1 + kP^2\right)^3}, \qquad F_{PQ} = \frac{h\left(1 - kP^2\right)}{\left(1 + kP^2\right)^2}, \qquad and \qquad F_{QQ} = 0$$

are the second partial derivatives of the function F.

(iv) *All adapted components of the* **curvature** *d-tensor* $\hat{\mathbf{R}}$ *of the canonical generalized Cartan connection* $C\hat{\Gamma}$, **produced by the cancer cell population flow (13.1)**, *cancel.*

(v) *The geometric* **electromagnetic distinguished** 2-**form, produced by the cancer cell population flow (13.1)**, *has the expression*

$$\hat{\mathbb{F}} = \hat{F}_{(i)j}^{(1)} \delta y_1^i \wedge dx^j,$$

where

$$\delta y_1^i = dy_1^i + \hat{N}_{(1)k}^{(i)} dx^k, \ \forall \ i = \overline{1,2},$$

and the adapted components $\hat{F}_{(i)j}^{(1)}$, $i,j = \overline{1,2}$, *are given by*

$$\hat{F}_{(1)1}^{(1)} = \hat{F}_{(2)2}^{(1)} = 0,$$

$$\hat{F}_{(2)1}^{(1)} = -\hat{F}_{(1)2}^{(1)} = \frac{1}{2} \left[(2a+1)P + aQ - (F_P + F_Q) \right]$$

$$= \frac{1}{2} \left[(2a+1)P + aQ - \frac{hQ\left(1-kP^2\right)}{\left(1+kP^2\right)^2} - \frac{hP}{1+kP^2} \right].$$

(vi) *The* **biologic geometric Yang-Mills energy, produced by the cancer cell population flow (13.1)**, *is given by the formula*

$$\mathcal{EYM}^{\mathrm{cancer}}(P,Q) = \frac{1}{4} \left[(2a+1)P + aQ - \frac{hQ\left(1-kP^2\right)}{\left(1+kP^2\right)^2} - \frac{hP}{1+kP^2} \right]^2.$$

Proof: We regard the cancer cell population flow (13.1) as a particular case of the first-order ODEs system (10.14) on the 1-jet space $J^1(\mathbb{R}, \mathbb{R}^2)$, with

$$n = 2, \ x^1 = P, \ x^2 = Q,$$

and

$$X_{(1)}^{(1)}(x^1, x^2) = x^1 - x^1(x^1 + x^2) + F(x^1, x^2),$$
$$X_{(1)}^{(2)}(x^1, x^2) = -rx^2 + ax^1(x^1 + x^2) - F(x^1, x^2).$$

Now, using Corollary 183, together with Remark 184, and taking into account that we have the Jacobian matrix

$$J\left(X_{(1)}\right) = \begin{pmatrix} 1 - 2P - Q + F_P & -P + F_Q \\ 2aP + aQ - F_P & -r + aP - F_Q \end{pmatrix}$$

$$= \begin{pmatrix} 1 - 2P - Q + \dfrac{hQ\left(1-kP^2\right)}{\left(1+kP^2\right)^2} & -P + \dfrac{hP}{1+kP^2} \\ 2aP + aQ - \dfrac{hQ\left(1-kP^2\right)}{\left(1+kP^2\right)^2} & -r + aP - \dfrac{hP}{1+kP^2} \end{pmatrix},$$

we obtain what we were looking for.

Open problem. The *Yang-Mills biologic energetic curves of constant level produced by the cancer cell population flow (13.1)*, which are different from the empty set, are, in the plane POQ, the curves of implicit equations

$$\Gamma_C : \left[(2a + 1) P + aQ - \frac{hQ\left(1 - kP^2\right)}{\left(1 + kP^2\right)^2} - \frac{hP}{1 + kP^2} \right]^2 = 4C,$$

where $C \geq 0$. For instance, the *zero Yang-Mills biologic energetic curve produced by the cancer cell population flow (13.1)* is, in the plane POQ, the graph of a rational function:

$$\Gamma_0 : Q = \frac{P\left(1 + kP^2\right)\left[h - (2a + 1)\left(1 + kP^2\right)\right]}{a\left(1 + kP^2\right)^2 - h\left(1 - kP^2\right)}.$$

As a possible opinion, we consider that if the cancer cell population flow does not generate any real Yang-Mills biologic energies, then it is to be expected that the variables P and Q vary along the rational curve Γ_0. Otherwise, if the cancer cell population flow generates a real Yang-Mills biologic energy, then it is possible that the shapes of the constant Yang-Mills biologic energetic curves Γ_C to offer useful interpretations for biologists.

13.2 THE JET RIEMANN-LAGRANGE GEOMETRY OF THE INFECTION BY HUMAN IMMUNODEFICIENCY VIRUS (HIV-1) EVOLUTION MODEL

The major target of HIV infection is a class of lymphocytes, or white blood cells, known as $CD4^+$ T cells. These cells secrete growth and differentiations factors that are required by other cell populations in the immune system, and hence these cells are also called "helper T cells." After becoming infected, the $CD4^+$ T cells can produce new HIV virus particles (or virions) so, in order to model HIV infection, it was introduced a population of uninfected target cells T, and productively infected cells T^*.

Over the past decade, a number of models have been developed to describe the immune system, its interaction with HIV, and the decline in $CD4^+$ T cells. We propose for our geometrical investigation a model that incorporates viral production (for more biologic details, see Perelson and Nelson [83]). This mathematical model of infection by HIV-1 relies on the variables $T(t)$ - *the population of uninfected target cells*, $T^*(t)$ - *the population of productively infected cells*, and $V(t)$ - *the HIV-1 virus*, whose evolution in time is given by the *HIV-1 flow* [83]

$$\begin{cases} \dfrac{dT}{dt} = s + (p - d)T - \dfrac{pT^2}{m} - kVT, \\[2mm] \dfrac{dT^*}{dt} = kTV - \delta T^*, \\[2mm] \dfrac{dV}{dt} = n\delta T^* - cV, \end{cases} \qquad (13.2)$$

where

- s represents the rate at which new T cells are created from sources within the body, such as thymus;

- p is the maximum proliferation rate of T cells;

- d is the death rate per T cells;

- δ represents the death rate for infected cells T^*;

- m is the T cells population density at which proliferation shuts off;

- k is the infection rate;

- n represents the total number of virions produced by a cell during its lifetime;

- c is the rate of clearance of virions.

In what follows, we apply our jet Riemann-Lagrange geometrical results to the *HIV-1 flow* (13.2) regarded on the 1-jet space $J^1(\mathbb{R}, \mathbb{R}^3)$. In this context, we obtain (cf. Corollary 183 and Remark 184):

Theorem 193. (i) *The* **canonical nonlinear connection** *on $J^1(\mathbb{R}, \mathbb{R}^3)$,* **produced by the HIV-1 flow (13.2)**, *has the local components*

$$\check{\Gamma} = \left(0, \check{N}^{(i)}_{(1)j} \right), \ i, j = \overline{1,3},$$

where $\check{N}^{(i)}_{(1)j}$ are the entries of the matrix

$$\check{N}_{(1)} = -\frac{1}{2} \begin{pmatrix} 0 & -kV & -kT \\ kV & 0 & kT - n\delta \\ kT & -kT + n\delta & 0 \end{pmatrix}.$$

(ii) *All adapted components of the* **canonical generalized Cartan connection** $C\check{\Gamma}$, **produced by the HIV-1 flow (13.2)**, *are zero.*

(iii) *All adapted components of the* **torsion** *d-tensor* $\check{\mathbf{T}}$ *of the canonical generalized Cartan connection* $C\check{\Gamma}$, **produced by the HIV-1 flow (13.2)**, *are zero, except the entries of the matrices*

$$\check{R}_{(1)1} = \begin{pmatrix} 0 & 0 & k/2 \\ 0 & 0 & -k/2 \\ -k/2 & k/2 & 0 \end{pmatrix}$$

and

$$\check{R}_{(1)3} = \begin{pmatrix} 0 & k/2 & 0 \\ -k/2 & 0 & 0 \\ 0 & 0 & 0 \end{pmatrix},$$

where

$$\check{R}_{(1)k} = \left(\check{R}^{(i)}_{(1)jk} \right)_{i,j=\overline{1,3}}, \ \forall \, k \in \{1, 3\}.$$

(iv) *All adapted components of the* **curvature** *d-tensor* $\check{\mathbf{R}}$ *of the canonical generalized Cartan connection* $C\check{\Gamma}$, **produced by the HIV-1 flow (13.2)**, *cancel.*

(v) *The geometric* **electromagnetic distinguished** 2-**form, produced by the HIV-1 flow (13.2)**, *has the expression*

$$\check{\mathbf{F}} = \check{F}^{(1)}_{(i)j} \delta y^i_1 \wedge dx^j,$$

where

$$\delta y^i_1 = dy^i_1 + \check{N}^{(i)}_{(1)k} dx^k, \ \forall \, i = \overline{1, 3},$$

and the adapted components $\check{F}^{(1)}_{(i)j}$ $i, j = \overline{1, 3}$, *are the entries of the matrix*

$$\check{F}^{(1)} = \frac{1}{2} \begin{pmatrix} 0 & -kV & -kT \\ kV & 0 & kT - n\delta \\ kT & -kT + n\delta & 0 \end{pmatrix}.$$

(vi) *The* **biologic geometric Yang-Mills energy, produced by the HIV-1 flow (13.2)**, *is given by the formula*

$$\mathcal{E}\mathcal{Y}\mathcal{M}^{\text{HIV-1}}(T, T^*, V) = \frac{1}{4} \left[k^2(V^2 + T^2) + (kT - n\delta)^2 \right].$$

Proof: Consider the HIV-1 flow (13.2) as a particular case of the first-order ODEs system (10.14) on the 1-jet space $J^1(\mathbb{R}, \mathbb{R}^3)$, with

$$n = 3, \ x^1 = T, \ x^2 = T^*, \ x^3 = V$$

and

$$X^{(1)}_{(1)}(x^1, x^2, x^3) = s + (p - d)x^1 - \frac{p}{m}(x^1)^2 - kx^3 x^1,$$

$$X^{(2)}_{(1)}(x^1, x^2, x^3) = kx^1 x^3 - \delta x^2,$$

$$X^{(3)}_{(1)}(x^1, x^2, x^3) = n\delta x^2 - cx^3.$$

It follows that we have the Jacobian matrix

$$J\left(X_{(1)} \right) = \begin{pmatrix} p - d - \dfrac{2p}{m}T - kV & 0 & -kT \\ kV & -\delta & kT \\ 0 & n\delta & -c \end{pmatrix}.$$

In conclusion, using Corollary 183 and Remark 184, we find the required result.

Open problem. The *Yang-Mills biologic energetic surfaces of constant level, produced by the HIV-1 flow (13.2)*, have in the system of axis OTT^*V the implicit equations

$$\Sigma_C : k^2(V^2 + T^2) + (kT - n\delta)^2 = 4C,$$

where $C \geq 0$. It is obvious that the surfaces Σ_C are some real or imaginary *cylinders*. Taking into account that the family of conics

$$\Gamma_C : 2k^2T^2 + k^2V^2 - 2kn\delta T + n^2\delta^2 - 4C = 0,$$

which generate the cylinders Σ_C, has the matrices

$$A = \begin{pmatrix} 2k^2 & 0 & -kn\delta \\ 0 & k^2 & 0 \\ -kn\delta & 0 & n^2\delta^2 - 4C \end{pmatrix},$$

it follows that its invariants are $\Delta_C = k^4 \left(n^2\delta^2 - 8C\right)$, $\delta = 2k^4 > 0$, and $I = 3k^2 > 0$. As a consequence, we have the following situations:

1. If $0 \leq C < \dfrac{n^2\delta^2}{8}$, then we have the *empty set* $\Sigma_{0 \leq C < \frac{n^2\delta^2}{8}} = \emptyset$.

2. If $C = \dfrac{n^2\delta^2}{8}$, then the surface $\Sigma_{C = \frac{n^2\delta^2}{8}}$ degenerates into the *straight line*

$$\Sigma_{C = \frac{n^2\delta^2}{8}} : \begin{cases} T = \dfrac{n\delta}{2k}, \\ V = 0. \end{cases}$$

3. If $C > \dfrac{n^2\delta^2}{8}$, then the surface $\Sigma_{C > \frac{n^2\delta^2}{8}}$ is a *right elliptic cylinder* of equation

$$\Sigma_{C > \frac{n^2\delta^2}{8}} : \dfrac{\left(T - \dfrac{n\delta}{2k}\right)^2}{a^2} + \dfrac{V^2}{b^2} = 1, \ T^* \in \mathbb{R},$$

where $a < b$ are given by

$$a = \dfrac{\sqrt{8C - n^2\delta^2}}{2k}, \ b = \dfrac{\sqrt{8C - n^2\delta^2}}{k\sqrt{2}}.$$

Obviously, it has as an axis of symmetry the straight line $\Sigma_{C = \frac{n^2\delta^2}{8}}$.

To what extent might the shapes of the above Yang-Mills energetic constant surfaces Σ_C provide valuable information for biologists?

13.3   FROM CALCIUM OSCILLATIONS ODEs SYSTEMS TO JET YANG-MILLS ENERGIES

Taking into account that we would like to develop the jet single-time Lagrange geometry of some ODEs systems that govern the calcium oscillations in a large variety of living cells, we would like to expose a few biological properties of these oscillations. So, we recall that the oscillations of cytosolic calcium concentration, known as calcium oscillations, play a vital role in providing the intracellular signaling and many biological processes are controlled by the oscillatory changes of cytosolic calcium concentration. Since the 1980's, when the self-sustained calcium oscillations were found experimentally by Woods, Cuthbertson, and Cobbold [100], many experimental works have been published.

Various models have been constructed to simulate calcium oscillations in living cells. In this Chapter, we will consider and geometrically study only two of these models. These mathematical models were proposed in the course of investigations of plausible mechanisms capable of generating complex calcium oscillations.

The first one was proposed by Borghans, Dupont and Goldbeter [23] and was deeply mathematically analyzed by Houart, Dupont, and Goldbeter [42]. This first mathematical model, describing the cytosolic calcium oscillations, relies on the interplay between CICR* (calcium-induced calcium release) and the Ca^{2+}-stimulated degradation of InsP$_3$ (inositol triphosphate).

Alternatively, the second mathematical model of calcium oscillations was introduced by Marhl, Haberichter, Brumen and Heinrich [50] and was also intensively studied from a mathematical point of view. This second model is based on the interplay between three calcium stores in the living cells: endoplasmic reticulum, mitochondria and cytosolic proteins.

In the following Subsections, we study the expressions of the abstract biological Yang-Mills energies produced by the calcium oscillations in this large variety of cell types. For reviews of these geometrical-biologic topics, see the works of Nicola [74], [68], and references therein.

13.3.1   Intracellular calcium oscillations induced by self-modulation of the inositol $1, 4, 5$-triphosphate signal

The mathematical model that describes calcium oscillations which is based on the mechanism of calcium-induced calcium release, takes into account the calcium-stimulated degradation of inositol triphosphate (InsP$_3$). This is because in some cell types, particularly in hepatocytes, calcium oscillations have been observed in response to stimulation by specific agonists. As these cells are not electrically excitable, it is likely that this calcium oscillations rely on the interplay between two intracellular mechanisms capable of destabilizing the steady state: An increase in InsP$_3$ is expected to lead to an increase in the frequency of calcium spikes, but at the same time the InsP$_3$-induced rise will also lead to increased InsP$_3$ hydrolysis due to the calcium activation of the InsP$_3$ 3-kinase.

The classical mathematical model for the study of cytosolic calcium oscillations and their associated degradations of InsP$_3$ in endoplasmic reticulum contains three variables $Z(t)$, $Y(t)$ and $A(t)$, where

- Z is the concentration of free calcium in the cytosol;
- Y is the concentration of free calcium in the internal pool;

- A is the InsP$_3$ concentration.

The time evolution of these variables is governed by the following first-order differential equations of cytosolic calcium oscillations, denoted by us with [Ca^{2+} − InsP$_3$] (for more biologic details, see Houart, Dupont, and Goldbeter [42]):

$$
\begin{cases}
\dfrac{dZ}{dt} = V_{M_3} \dfrac{Z^m}{K_Z^m + Z^m} \cdot \dfrac{Y^2}{K_Y^2 + Y^2} \cdot \dfrac{A^4}{K_A^4 + A^4} - V_{M_2} \dfrac{Z^2}{K_2^2 + Z^2} \\[2ex]
\qquad + k_f Y - kZ + V_0 + \beta V_1, \\[2ex]
\dfrac{dY}{dt} = V_{M_2} \dfrac{Z^2}{K_2^2 + Z^2} - k_f Y - V_{M_3} \dfrac{Z^m}{K_Z^m + Z^m} \cdot \dfrac{Y^2}{K_Y^2 + Y^2} \cdot \dfrac{A^4}{K_A^4 + A^4}, \\[2ex]
\dfrac{dA}{dt} = \beta V_{M_4} - V_{M_5} \dfrac{A^p}{K_5^p + A^p} \cdot \dfrac{Z^n}{K_d^n + Z^n} - \varepsilon A,
\end{cases}
$$

where

- V_0 refers to a constant input of calcium from the extracellular medium;

- V_1 is the maximum rate of stimulus-induced influx of calcium from the extracellular medium;

- β is a constant parameter reflecting the degree of stimulation of the cell by an agonist and thus only varies between 0 and 1;

- the rates $V_2 = V_{M_2} \frac{Z^2}{K_2^2 + Z^2}$ and $V_3 = V_{M_3} \frac{Z^m}{K_Z^m + Z^m} \cdot \frac{Y^2}{K_Y^2 + Y^2} \cdot \frac{A^4}{K_A^4 + A^4}$ refer to the pumping of cytosolic calcium into the internal stores and to the release of calcium from these stores into the cytosol in a process activated by cytosolic calcium, respectively. The constants V_{M_2} and V_{M_3} represent the maximum values of the preceding rates;

- the parameters K_2, K_Y, K_Z, and K_A are threshold constants for pumping, release, and activation of release by calcium and by InsP$_3$;

- k_f is a rate constant measuring the passive, linear leak of Y into Z;

- k relates on the assumed linear transport of cytosolic calcium into the extracellular medium;

- V_{M_4} is the maximum rate of stimulus-induced synthesis of InsP$_3$;

- $V_5 = V_{M_5} \frac{A^p}{K_5^p + A^p} \cdot \frac{Z^n}{K_d^n + Z^n}$ is the rate of phosphorylation of the InsP$_3$ by the 3-kinase, which is characterized by a maximum value V_{M_5} and a half-saturation constant K_5;

- m, n, and p are the Hill's coefficients related to the cooperative processes;

- ε is the rate of phosphorylation of the InsP$_3$ by the 5-phosphatase.

From a biological point of view, we recall that the preceding ODEs system is based on the mechanism of Ca^{2+}-induced Ca^{2+} release (CICR), that takes into account the Ca^{2+} stimulates the degradations of the inositol $1, 4, 5$ triphosphate $InsP_3$ by a 3-kinase.

From a geometrical point of view, the ODEs system of calcium oscillations and the degradation of the inositol trisphosphate $InsP_3$ may be regarded as an ODEs system on the particular 1-jet space $J^1(\mathbb{R}, \mathbb{R}^3)$. Let us denote the coordinates of the spatial manifold $\mathbb{R}^3$ by $x^1 = Z$, $x^2 = Y$, and $x^3 = A$. Consequently, the local coordinates on the 1-jet space $J^1(\mathbb{R}, \mathbb{R}^3)$ are

$$(t, \ x^1 = Z, \ x^2 = Y, \ x^3 = A, \ y_1^1 = \dot{Z}, \ y_1^2 = \dot{Y}, \ y_1^3 = \dot{A}).$$

In this geometrical context, the ODEs system $[Ca^{2+} - InsP_3]$, as a particular ODEs system of the form (10.14), is determined by the following distinguished tensorial components $X^{(i)}_{(1)}(x^1, x^2, x^3)$:

$$X^{(1)}_{(1)}(Z, Y, A) = \ V_{M_3} \frac{Z^m}{K_Z^m + Z^m} \cdot \frac{Y^2}{K_Y^2 + Y^2} \cdot \frac{A^4}{K_A^4 + A^4} - V_{M_2} \frac{Z^2}{K_2^2 + Z^2}$$

$$+ V_0 + \beta V_1 + k_f Y - k Z,$$

$$X^{(2)}_{(1)}(Z, Y, A) = \ V_{M_2} \frac{Z^2}{K_2^2 + Z^2} - k_f Y - V_{M_3} \frac{Z^m}{K_Z^m + Z^m} \cdot \frac{Y^2}{K_Y^2 + Y^2} \cdot \frac{A^4}{K_A^4 + A^4},$$

$$X^{(3)}_{(1)}(Z, Y, A) = \ \beta V_{M_4} - V_{M_5} \frac{A^p}{K_5^p + A^p} \cdot \frac{Z^n}{K_d^n + Z^n} - \varepsilon A.$$

Consequently, using the general results from Chapter 10, together with some partial derivatives and computations, we find the following important geometrical result that characterizes the cytosolic calcium oscillations in hepatocytes and the degradation of $InsP_3$ through endoplasmic reticulum (cf. Corollary 183):

Theorem 194. *The **biologic electromagnetic field** $\mathbb{F}$ produced by the ODEs system $[Ca^{2+}\text{-}InsP_3]$ and the pair of Euclidian metrics $\Delta = (1, \delta_{ij})$ has the components*

$$F^{(1)}_{(1)2} = \ \frac{1}{2} \left\{ k_f - 2 V_{M_2} K_2^2 \frac{Z}{(K_2^2 + Z^2)^2} + V_{M_3} \frac{Z^{m-1}}{K_Z^m + Z^m} \cdot \frac{Y}{K_Y^2 + Y^2} \cdot \frac{A^4}{K_A^4 + A^4} \right.$$

$$\left. \cdot \left[2 \frac{K_Y^2 Z}{K_Y^2 + Y^2} + m \frac{K_Z^m Y}{K_Z^m + Z^m} \right] \right\},$$

$$F^{(1)}_{(1)3} = \ \frac{2 V_{M_3} K_A^4 Z^m}{K_Z^m + Z^m} \cdot \frac{Y^2}{K_Y^2 + Y^2} \cdot \frac{A^3}{(K_A^4 + A^4)^2} + \frac{n}{2} \cdot \frac{A^p}{K_5^p + A^p} \cdot \frac{V_{M_5} K_d^n Z^{n-1}}{(K_d^n + Z^n)^2},$$

$$F^{(1)}_{(2)3} = \ -2 V_{M_3} K_A^4 \cdot \frac{Z^m}{K_Z^m + Z^m} \cdot \frac{Y^2}{K_Y^2 + Y^2} \cdot \frac{A^3}{(K_A^4 + A^4)^2}.$$

Proof: We have the formulas

$$F^{(1)}_{(1)2} = \frac{1}{2}\left[\frac{\partial X^{(1)}_{(1)}}{\partial Y} - \frac{\partial X^{(2)}_{(1)}}{\partial Z}\right], \quad F^{(1)}_{(1)3} = \frac{1}{2}\left[\frac{\partial X^{(1)}_{(1)}}{\partial A} - \frac{\partial X^{(3)}_{(1)}}{\partial Z}\right],$$

$$F^{(1)}_{(2)3} = \frac{1}{2}\left[\frac{\partial X^{(2)}_{(1)}}{\partial A} - \frac{\partial X^{(3)}_{(1)}}{\partial Y}\right].$$

So, some direct computations imply the required result.

We recall that the formulas from Theorem 170 show that the spatial components $G^{(i)}_{(1)1}$ of the semispray $\mathcal{S}_{[\mathrm{Ca}^{2+}\text{-InsP}_3]} = \left(0, G^{(i)}_{(1)1}\right)$, produced by the ODEs system $[\mathrm{Ca}^{2+}\text{-InsP}_3]$ and the pair of Euclidian metrics $\Delta = (1, \delta_{ij})$, are given by

$$G^{(i)}_{(1)1} = -\frac{1}{2}\left\{\frac{\partial X^{(r)}_{(1)}}{\partial x^i}\left[X^{(r)}_{(1)} - y^r_1\right] + \frac{\partial X^{(i)}_{(1)}}{\partial x^m}y^m_1\right\}.$$

Consequently, Theorem 170 implies the following interesting qualitative geometrical result with biological-energetic connotations:

Theorem 195. *The solutions of class C^2 of the ODEs system $[\mathrm{Ca}^{2+}\text{-InsP}_3]$ may be regarded as harmonic curves of the semispray $\mathcal{S}_{[\mathrm{Ca}^{2+}\text{-InsP}_3]}$ on the 1-jet space $J^1(\mathbb{R}, \mathbb{R}^3)$. In other words, the C^2 solutions $(Z(t), Y(t), A(t))$ of the first-order ODEs system $[\mathrm{Ca}^{2+}\text{-InsP}_3]$ satisfy also the second-order biological ODEs system:*

$$\begin{cases} \dfrac{d^2 Z}{dt^2} - 2F^{(1)}_{(1)2}\dfrac{dY}{dt} - 2F^{(1)}_{(1)3}\dfrac{dA}{dt} - \displaystyle\sum_{k=1}^{3}\partial_Z X^{(k)}_{(1)} \cdot X^{(k)}_{(1)} = 0, \\[4mm] \dfrac{d^2 Y}{dt^2} + 2F^{(1)}_{(1)2}\dfrac{dZ}{dt} - 2F^{(1)}_{(2)3}\dfrac{dA}{dt} - \displaystyle\sum_{k=1}^{3}\partial_Y X^{(k)}_{(1)} \cdot X^{(k)}_{(1)} = 0, \\[4mm] \dfrac{d^2 A}{dt^2} + 2F^{(1)}_{(1)3}\dfrac{dZ}{dt} + 2F^{(1)}_{(2)3}\dfrac{dY}{dt} - \displaystyle\sum_{k=1}^{3}\partial_A X^{(k)}_{(1)} \cdot X^{(k)}_{(1)} = 0. \end{cases} \quad (13.3)$$

Proof: We use the general second-order differential equations of the harmonic curves of a semispray, which are given by (1.12). We take also into account that we have $x^1 = Z$, $x^2 = Y$, and $x^3 = A$, together with the particular semispray $\mathcal{S}_{[\mathrm{Ca}^{2+}\text{-InsP}_3]}$.

Remark 196. The importance of the second-order system (13.3) is that its equations are equivalent with the Euler-Lagrange equations of the least squares Lagrangian on the 1-jet space $J^1(\mathbb{R}, \mathbb{R}^3)$, given by

$$\mathcal{LSQ}^{[\mathrm{Ca}^{2+}\text{-InsP}_3]} = \left(\dot{Z} - X^{(1)}_{(1)}(Z, Y, A)\right)^2 + \left(\dot{Y} - X^{(2)}_{(1)}(Z, Y, A)\right)^2 + \left(\dot{A} - X^{(3)}_{(1)}(Z, Y, A)\right)^2.$$

Therefore, the solutions $(Z(t), Y(t), A(t))$ of class C^2 of the first-order ODEs system $[Ca^{2+}\text{-}InsP_3]$ are global minimum points for the jet least squares biological Lagrangian $\mathcal{LSQ}^{[Ca^{2+}\text{-}InsP_3]}$.

Particularizing the general definition of the geometric jet Yang-Mills energy of a general first-order ODEs system to our present biological phenomena, we deduce that the *biological Yang-Mills energy* produced by the intracellular calcium oscillations in some non-excitable cell types has the form (cf. Definition 181)

$$\mathcal{EYM}_{[Ca^{2+}\text{-}InsP_3]}(Z, Y, A) = \left[F^{(1)}_{(1)2}\right]^2 + \left[F^{(1)}_{(1)3}\right]^2 + \left[F^{(1)}_{(2)3}\right]^2.$$

It is important to note that we have three main sets of parameter values for $[Ca^{2+}\text{-}InsP_3]$, including *bursting*, *chaos* and *quasiperiodicity* situations, which are listed in the following table (for more biologic details, see Houart, Dupont, and Goldbeter [42]):

Parameters	Bursting	Chaos	Quasiperiodicity
β	0.46	0.65	0.51
n	2	4	4
m	4	2	2
p	1	1	2
K_2 (μM)	0.1	0.1	0.1
K_5 (μM)	1	0.3194	0.3
K_A (μM)	0.1	0.1	0.2
K_d (μM)	0.6	1	0.5
K_Y (μM)	0.2	0.3	0.2
K_Z (μM)	0.3	0.6	0.5
k (s^{-1})	0.1667	0.1667	0.1667
k_f (s^{-1})	0.0167	0.0167	0.0167
ε (s^{-1})	0.0167	0.2167	0.0017
V_0 (μM s^{-1})	0.0333	0.0333	0.0333
V_1 (μM s^{-1})	0.0333	0.0333	0.0333
V_{M_2} (μM s^{-1})	0.1	0.1	0.1
V_{M_3} (μM s^{-1})	0.3333	0.5	0.3333
V_{M_4} (μM s^{-1})	0.0417	0.05	0.0833
V_{M_5} (μM s^{-1})	0.5	0.8333	0.5

These parameter values correspond to the various types of complex oscillatory behavior observed in the model defined by equations $[\text{Ca}^{2+}\text{-InsP}_3]$ and obviously produce particular and distinct geometric biological Yang-Mills energies:

Theorem 197. *The following formulas for the geometric biological energies of Yang-Mills type are true:*

(i) *Biological Yang-Mills energy of* **bursting** *cytosolic calcium oscillations in the model involving* Ca^{2+} *activated* InsP_3 *degradation.*

$$
\mathcal{EYM}^{\text{bursting}}_{[\text{Ca}^{2+}\text{-InsP}_3]} = \left\{ 0.00835 - \frac{0.001 \cdot Z}{(0.01 + Z^2)^2} + \frac{0.3333 \cdot Z^3}{0.0081 + Z^4} \cdot \frac{Y}{0.04 + Y^2} \cdot \frac{A^4}{0.0001 + A^4} \right.
$$

$$
\cdot \left[\frac{0.04 \cdot Z}{0.04 + Y^2} + \frac{0.0162 \cdot Y}{0.0081 + Z^4} \right] \right\}^2 + \left\{ \frac{0.6666 \cdot 10^{-4} \cdot Z^4}{0.0081 + Z^4} \cdot \frac{Y^2}{0.04 + Y^2} \cdot \frac{A^3}{(0.0001 + A^4)^2} \right.
$$

$$
+ \frac{A}{(1 + A)^2} \cdot \frac{0.18 \cdot Z}{(0.36 + Z^2)^2} \right\}^2 + \frac{0.44435 \cdot 10^{-8} \cdot Z^8}{(0.0081 + Z^4)^2} \cdot \frac{Y^4}{(0.04 + Y^2)^2} \cdot \frac{A^6}{(0.0001 + A^4)^4}.
$$

(ii) *Biological Yang-Mills energy of* **chaos** *cytosolic calcium oscillations in the model involving* Ca^{2+} *activated* InsP_3 *degradation.*

$$
\mathcal{EYM}^{\text{chaos}}_{[\text{Ca}^{2+}\text{-InsP}_3]} = \left\{ 0.00835 - \frac{0.001 \cdot Z}{(0.01 + Z^2)^2} + \frac{0.5 \cdot Z}{0.36 + Z^2} \cdot \frac{Y}{0.09 + Y^2} \cdot \frac{A^4}{0.0001 + A^4} \right.
$$

$$
\cdot \left[\frac{0.09 \cdot Z}{0.09 + Y^2} + \frac{0.36 \cdot Y}{0.36 + Z^2} \right] \right\}^2 + \left\{ \frac{0.0001 \cdot Z^2}{0.36 + Z^2} \cdot \frac{Y^2}{0.09 + Y^2} \cdot \frac{A^3}{(0.0001 + A^4)^2} \right.
$$

$$
+ \frac{2A}{0.3194 + A} \cdot \frac{0.8333 \cdot Z^3}{(1 + Z^4)^2} \right\}^2 + \frac{10^{-8} \cdot Z^4}{(0.36 + Z^2)^2} \cdot \frac{Y^4}{(0.09 + Y^2)^2} \cdot \frac{A^6}{(0.0001 + A^4)^4};
$$

(iii) *Biological Yang-Mills energy of* **quasiperiodicity** *cytosolic calcium oscillations in the model involving* Ca^{2+} *activated* InsP_3 *degradation.*

$$
\mathcal{EYM}^{\text{quasiperiodicity}}_{[\text{Ca}^{2+}\text{-InsP}_3]} = \left\{ 0.00835 - \frac{0.001 \cdot Z}{(0.01 + Z^2)^2} + \frac{0.3333 \cdot Z}{0.25 + Z^2} \cdot \frac{Y}{0.04 + Y^2} \cdot \frac{A^4}{0.0016 + A^4} \right.
$$

$$
\cdot \left[\frac{0.04 \cdot Z}{0.09 + Y^2} + \frac{0.25 \cdot Y}{0.25 + Z^2} \right] \right\}^2 + \left\{ \frac{0.00106656 \cdot Z^2}{0.25 + Z^2} \cdot \frac{Y^2}{0.04 + Y^2} \cdot \frac{A^3}{(0.0016 + A^4)^2} \right.
$$

$$
+ \frac{2A^2}{0.09 + A^2} \cdot \frac{0.03125 \cdot Z^3}{(0.0625 + Z^4)^2} \right\}^2 + \frac{0.0000113 \cdot Z^4}{(0.25 + Z^2)^2} \cdot \frac{Y^4}{(0.04 + Y^2)^2} \cdot \frac{A^6}{(0.0016 + A^4)^4}.
$$

13.3.2 Calcium oscillations in a model involving endoplasmic reticulum, mitochondria, and cytosolic proteins

The next mathematical model (that we will geometrically study in the sequel) represents a possible mechanism for complex calcium oscillations based on the interplay between three calcium stores in the biological living cells: the endoplasmic reticulum (ER), mitochondria, and cytosolic proteins. The majority of calcium released from the ER is first very quickly sequestred by mitochondria. Afterward, a much slower release of calcium from the mitochondria serves as the calcium supply for the intermediate calcium exchanges between the ER and the cytosolic proteins. We would like to point out that the oscillations of cytosolic calcium concentration play a vital role in providing the intracellular signaling. Moreover, many cellular processes, like cell secretion or egg fertilization, for instance, are controlled by the oscillatory regime of the cytosolic calcium concentration.

In this second mathematical model, we have three variables $Ca_{cyt}(t)$, $Ca_{ER}(t)$, and $Ca_m(t)$, where

- Ca_{cyt} means the free cytosolic calcium concentration;

- Ca_{ER} means the free calcium concentration in the ER;

- Ca_m means the free calcium concentration in the mitochondria.

The preceding variables of calcium are governed by the following first-order ODEs system of the calcium oscillations through endoplasmic reticulum, mitochondria and the cytosolic proteins, denoted by us as $[Ca^{2+}\text{-}ER\text{-}cyt.pr\text{-}m]$ (for more details, see Marhl, Haberichter, Brumen, and Heinrich [50]):

$$
\begin{cases}
\dfrac{dCa_{cyt}}{dt} = k_{ch}\dfrac{Ca_{cyt}^2}{K_1^2 + Ca_{cyt}^2}(Ca_{ER} - Ca_{cyt}) + k_{leak}(Ca_{ER} - Ca_{cyt}) \\[2mm]
\qquad - k_{pump}Ca_{cyt} + \left(k_{out}\dfrac{Ca_{cyt}^2}{K_1^2 + Ca_{cyt}^2} + k_m\right)Ca_m \\[2mm]
\qquad - k_{in}\dfrac{Ca_{cyt}^8}{K_2^8 + Ca_{cyt}^8} + k_-\left(Ca_{tot} - Ca_{cyt} - \dfrac{\rho_{ER}}{\beta_{ER}}Ca_{ER}\right) \\[2mm]
\qquad - dfrac\rho_m\beta_m Ca_m) - k_+ Ca_{cyt}(Pr_{tot} - Ca_{tot} + Ca_{cyt} \\[2mm]
\qquad + \dfrac{\rho_{ER}}{\beta_{ER}}Ca_{ER} + \dfrac{\rho_m}{\beta_m}Ca_m\Big), \\[3mm]
\dfrac{dCa_{ER}}{dt} = \dfrac{\beta_{ER}}{\rho_{ER}}\left[k_{pump}Ca_{cyt} - k_{ch}\dfrac{Ca_{cyt}^2}{K_1^2 + Ca_{cyt}^2}(Ca_{ER} - Ca_{cyt})\right. \\[2mm]
\qquad \left. - k_{leak}(Ca_{ER} - Ca_{cyt})\right] \\[3mm]
\dfrac{dCa_m}{dt} = \dfrac{\beta_m}{\rho_m}\left[k_{in}\dfrac{Ca_{cyt}^8}{K_2^8 + Ca_{cyt}^8} - \left(k_{out}\dfrac{Ca_{cyt}^2}{K_1^2 + Ca_{cyt}^2} + k_m\right)Ca_m\right],
\end{cases}
$$

where

- Pr_{tot} is the total concentration of cytosolic proteins;

- Ca_{tot} represents the total cellular Ca^{2+} concentration;

- K_1 represents the half-saturation for Ca^{2+};

- K_2 represents the half-saturation for Ca^{2+} of uniporters in the mitochondrial membrane;

- $V_{pump} = k_{pump}Ca_{cyt}$ is the adenosine triphosphate (ATP)-dependent calcium uptake from the cytosol into the ER;

- $V_{ch} = k_{ch}\dfrac{Ca_{cyt}^2}{K_1^2 + Ca_{cyt}^2}(Ca_{ER} - Ca_{cyt})$ is the calcium efflux from the ER through channels following the calcium-induced calcium release mechanism;

- $V_{leak} = k_{leak}(Ca_{ER} - Ca_{cyt})$ represents the calcium leak flux from the ER into the cytosol;

- $V_{in} = k_{in}\dfrac{Ca_{cyt}^8}{K_2^8 + Ca_{cyt}^8}$ is the active calcium uptake by mitochondrial uniporters;

- $V_{out} = \left(k_{out}\dfrac{Ca_{cyt}^2}{K_1^2 + Ca_{cyt}^2} + k_m\right)Ca_m$ is a very small non-specific leak flux;

- k_- and k_+ denote the off and the on rate constants of the calcium binding;

- ρ_{ER} and ρ_m represent the volume ratio between the ER and the cytosol or between the mitochondria and the cytosol, respectively;

- β_{ER} and β_m are constant factors for relating the concentrations of free calcium in the ER and the mitochondria to the respective total concentrations;

- k_{pump} is the rate constant of the ATP;

- k_{ch} represents the maximal permeability of the calcium channels in the ER membrane;

- k_{leak} is the rate constant for calcium leak flux through the ER membrane;

- k_{in} represents the maximal permeability of the uniporters in the mitochondrial membrane;

- k_{out} represents the maximal rate for calcium flux through pores;

- k_m stands for the non-specific leak flux;

Remark 198. From a biological point of view, in addition to the endoplasmic reticulum as the main intracellular calcium store used in the first mathematical model, in this second model, the mitochondrial and cytosolic Ca^{2+}-binding proteins are also taken into account. We recall that this model was proposed in [50] especially for the study of the physiological role of mitochondria and the cytosolic proteins in generating complex Ca^{2+} oscillations. For more biological details, please see the works [68] and [74] and references therein.

From a differential geometric point of view, we underline that the first-order ODEs system of calcium oscillations through endoplasmic reticulum, mitochondria and cytosolic proteins may be regarded as an ODEs system on the particular 1-jet space $J^1(\mathbb{R}, \mathbb{R}^3)$. Denoting the coordinates of the manifold $\mathbb{R}^3$ by $x^1 = Ca_{cyt}$, $x^2 = Ca_{ER}$ and $x^3 = Ca_m$, it follows that the local coordinates on the 1-jet space $J^1(\mathbb{R}, \mathbb{R}^3)$ are

$$(t,\ x^1 = Ca_{cyt},\ x^2 = Ca_{ER},\ x^3 = Ca_m,\ y_1^1 = \dot{Ca}_{cyt},\ y_1^2 = \dot{Ca}_{ER},\ y_1^3 = \dot{Ca}_m).$$

We also remark that the ODEs system of order one $[Ca^{2+}\text{-}ER\text{-}cyt.pr\text{-}m]$ is determined by the following distinguished tensorial components $X_{(1)}^{(i)}(x^1, x^2, x^3)$:

$$X_{(1)}^{(1)}(Ca_{cyt}, Ca_{ER}, Ca_m) = \frac{k_{ch}Ca_{cyt}^2}{K_1^2 + Ca_{cyt}^2}(Ca_{ER} - Ca_{cyt}) + k_{leak}(Ca_{ER} - Ca_{cyt})$$

$$-k_{pump}Ca_{cyt} + \left(k_{out}\frac{Ca_{cyt}^2}{K_1^2 + Ca_{cyt}^2} + k_m\right)Ca_m - k_{in}\frac{Ca_{cyt}^8}{K_2^8 + Ca_{cyt}^8}$$

$$+k_-\left(Ca_{tot} - Ca_{cyt} - \frac{\rho_{ER}}{\beta_{ER}}Ca_{ER} - \frac{\rho_m}{\beta_m}Ca_m\right)$$

$$-k_+Ca_{cyt}\left(Pr_{tot} - Ca_{tot} + Ca_{cyt} + \frac{\rho_{ER}}{\beta_{ER}}Ca_{ER} + \frac{\rho_m}{\beta_m}Ca_m\right),$$

$$X_{(1)}^{(2)}(Ca_{cyt}, Ca_{ER}, Ca_m) = \frac{\beta_{ER}}{\rho_{ER}}\left[k_{pump}Ca_{cyt} - k_{leak}(Ca_{ER} - Ca_{cyt})\right.$$

$$\left. - k_{ch}\frac{Ca_{cyt}^2}{K_1^2 + Ca_{cyt}^2}(Ca_{ER} - Ca_{cyt})\right],$$

$$X_{(1)}^{(3)}(Ca_{cyt}, Ca_{ER}, Ca_m) = \frac{\beta_m}{\rho_m}\left[\frac{k_{in}Ca_{cyt}^8}{K_2^8 + Ca_{cyt}^8} - \left(k_{out}\frac{Ca_{cyt}^2}{K_1^2 + Ca_{cyt}^2} + k_m\right)Ca_m\right].$$

As in the preceding biological case, it is obvious that again some partial derivatives and computations imply some geometrical results which, in our opinion, are possible to characterize the microscopic energetic changes produced by the calcium oscillations in the model involving endoplasmic reticulum, mitochondria, and cytosolic proteins (cf. Corollary 183):

Theorem 199. *The adapted components of the* **biologic electromagnetic field** $\mathbb{F}$ *produced by the ODEs system* $[Ca^{2+}\text{-}ER\text{-}cyt.pr\text{-}m]$ *and the pair of Euclidian metrics* $\Delta = (1, \delta_{ij})$ *are given by the following expressions:*

$$F_{(1)2}^{(1)} = \frac{1}{2} \left[\frac{\partial X_{(1)}^{(1)}}{\partial Ca_{ER}} - \frac{\partial X_{(1)}^{(2)}}{\partial Ca_{cyt}} \right] = \frac{1}{2} \left\{ k_{ch} \frac{Ca_{cyt}^2}{K_1^2 + Ca_{cyt}^2} + k_{leak} - \frac{\beta_{ER}}{\rho_{ER}} \right.$$

$$\cdot \left[k_{pump} - \frac{2 k_{ch} K_1^2 Ca_{cyt}}{(K_1^2 + Ca_{cyt}^2)^2} (Ca_{ER} - Ca_{cyt}) + k_{ch} \frac{Ca_{cyt}^2}{K_1^2 + Ca_{cyt}^2} + k_{leak} \right]$$

$$\left. - \frac{\rho_{ER}}{\beta_{ER}} (k_- + k_+ Ca_{cyt}) \right\},$$

$$F_{(1)3}^{(1)} = \frac{1}{2} \left[\frac{\partial X_{(1)}^{(1)}}{\partial Ca_m} - \frac{\partial X_{(1)}^{(3)}}{\partial Ca_{cyt}} \right] = \frac{1}{2} \left\{ k_{out} \frac{Ca_{cyt}^2}{K_1^2 + Ca_{cyt}^2} + k_m - \frac{\beta_m}{\rho_m} \right.$$

$$\cdot \left[8 k_{in} K_2^8 \frac{Ca_{cyt}^7}{(K_2^8 + Ca_{cyt}^8)^2} - 2 k_{out} K_1^2 \frac{Ca_m Ca_{cyt}}{(K_1^2 + Ca_{cyt}^2)^2} \right]$$

$$\left. - \frac{\rho_m}{\beta_m} (k_- + k_+ Ca_{cyt}) \right\},$$

$$F_{(2)3}^{(1)} = \frac{1}{2} \left[\frac{\partial X_{(1)}^{(2)}}{\partial Ca_m} - \frac{\partial X_{(1)}^{(3)}}{\partial Ca_{ER}} \right] = 0.$$

Taking again into account that the spatial components $G_{(1)1}^{(i)}$ of the semispray

$$\mathcal{S}_{[Ca^{2+}-ER\text{-}cyt.pr\text{-}m]} = \left(0, G_{(1)1}^{(i)} \right),$$

produced by the ODEs system $[Ca^{2+} - ER - cyt.pr - m]$ and the pair of Euclidian metrics $\Delta = (1, \delta_{ij})$, are given by

$$G_{(1)1}^{(i)} = -\frac{1}{2} \left\{ \frac{\partial X_{(1)}^{(m)}}{\partial x^i} X_{(1)}^{(m)} + \left(\frac{\partial X_{(1)}^{(i)}}{\partial x^m} - \frac{\partial X_{(1)}^{(m)}}{\partial x^i} \right) y_1^m \right\},$$

we naturally establish the following qualitative geometrical result with biological-energetic meaning:

Theorem 200. *The C^2 solutions of the ODEs system $[Ca^{2+}\text{-}ER\text{-}cyt.pr\text{-}m]$ may be regarded as harmonic curves of the semispray $\mathcal{S}_{[Ca^{2+}\text{-}ER\text{-}cyt.pr\text{-}m]}$ on the 1-jet space $J^1(\mathbb{R}, \mathbb{R}^3)$. In other words, the solutions $(Ca_{cyt}(t), Ca_{ER}(t), Ca_m(t))$ of the first-order ODEs system $[Ca^{2+}\text{-}ER\text{-}cyt.pr\text{-}m]$ verify the second-order biological ODEs system*

$$\begin{cases} \dfrac{d^2\text{Ca}_{cyt}}{dt^2} - 2F^{(1)}_{(1)2}\dfrac{d\text{Ca}_{ER}}{dt} - 2F^{(1)}_{(1)3}\dfrac{d\text{Ca}_m}{dt} - \displaystyle\sum_{k=1}^{3} X^{(k)}_{(1)} \cdot \partial_{\text{Ca}_{cyt}} X^{(k)}_{(1)} = 0, \\[2em] \dfrac{d^2\text{Ca}_{ER}}{dt^2} + 2F^{(1)}_{(1)2}\dfrac{d\text{Ca}_{cyt}}{dt} - \displaystyle\sum_{k=1}^{3} X^{(k)}_{(1)} \cdot \partial_{\text{Ca}_{ER}} X^{(k)}_{(1)} = 0, \\[2em] \dfrac{d^2\text{Ca}_m}{dt^2} + 2F^{(1)}_{(1)3}\dfrac{d\text{Ca}_{cyt}}{dt} - \displaystyle\sum_{k=1}^{3} X^{(k)}_{(1)} \cdot \partial_{\text{Ca}_m} X^{(k)}_{(1)} = 0. \end{cases}$$

$$(13.4)$$

Proof: We use the general second-order differential equations of the harmonic curves of a semispray, which are given by (1.12). At the same time, we take into account that we have $x^1 = \text{Ca}_{cyt}$, $x^2 = \text{Ca}_{ER}$, and $x^3 = \text{Ca}_m$, together with the particular semispray $\mathcal{S}_{[\text{Ca}^{2+}\text{-}ER\text{-}cyt.pr\text{-}m]}$.

Remark 201. The second-order differential equations (13.4) are obviously equivalent with the Euler-Lagrange equations of the jet least squares Lagrangian

$$\mathcal{LSQ}^{[\text{Ca}^{2+}-ER-cyt.pr-m]} = \left(\dot{\text{Ca}}_{cyt} - X^{(1)}_{(1)}\right)^2 + \left(\dot{\text{Ca}}_{ER} - X^{(2)}_{(1)}\right)^2 \\ + \left(\dot{\text{Ca}}_m - X^{(3)}_{(1)}\right)^2.$$

In other words, the C^2 solutions $(\text{Ca}_{cyt}(t), \text{Ca}_{ER}(t), \text{Ca}_m(t))$ of the first-order ODEs system $[\text{Ca}^{2+}\text{-}ER\text{-}cyt.pr\text{-}m]$ are global minimum points for the jet least squares Lagrangian $\mathcal{LSQ}^{[\text{Ca}^{2+}\text{-}ER\text{-}cyt.pr\text{-}m]}$.

In the sequel, particularizing the general definition of the geometric Yang-Mills energy of a general ODEs system to our present biological phenomena, we deduce that the biological Yang-Mills energy produced by the calcium oscillations in this model (that takes into account endoplasmic reticulum, mitochondria, and cytosolic proteins) has the form (cf. Definition 181)

$$\mathcal{EYM}_{[\text{Ca}^{2+}-ER-cyt.pr-m]}(\text{Ca}_{cyt}, \text{Ca}_{ER}, \text{Ca}_m) = \left[F^{(1)}_{(1)2}\right]^2 + \left[F^{(1)}_{(1)3}\right]^2,$$

because $F^{(1)}_{(2)3} = 0$.

Also, it is important to note that we have three sets of parameter values (corresponding to three types of complex Ca^{2+} oscillations in which the parameter values correspond to the various types of complex oscillatory whose behavior was observed in the model $[\text{Ca}^{2+}\text{-}ER\text{-}cyt.pr\text{-}m]$), including *bursting*, *chaos*, and *birhythmicity* situations, which are listed in the following table (for more biologic details, see Marhl, Haberichter, Brumen and Heinrich [50]):

Parameters	Bursting	Chaos	Birhythmicity
Ca_{tot} (μM)	90	90	90
Pr_{tot} (μM)	120	120	120
ρ_{ER}	0.01	0.01	0.01
ρ_m	0.01	0.01	0.01
β_{ER}	0.0025	0.0025	0.0025
β_m	0.0025	0.0025	0.0025
K_1 (μM)	5	5	5
K_2 (μM)	0.8	0.8	0.8
k_{ch} (s^{-1})	4100	2780 - 2980, 3598 - 3636	1968-2456
k_{pump} (s^{-1})	20	20	20
k_{leak} (s^{-1})	0.05	0.05	0.05
k_{in} (μM) (s^{-1})	300	300	300
k_{out} (s^{-1})	125	125	125
k_m (s^{-1})	0.00625	0.00625	0.00625
k_+ (μM) (s^{-1})	0.1	0.1	0.1
k_- (s^{-1})	0.01	0.01	0.01

Obviously the preceding parameter values produce particular and distinct geometric biological Yang-Mills energies for calcium oscillations phenomena through endoplasmic reticulum, mitochondria, and cytosolic proteins.

Theorem 202. *The following formulas for the geometric biological energies of Yang-Mills type are true:*

(i) *Biological Yang-Mills energy of* **bursting** *calcium oscillations in the model involving endoplasmic reticulum, mitochondria, and cytosolic proteins.*

$$\mathcal{E}\mathcal{Y}\mathcal{M}^{\text{bursting}}_{[Ca^{2+}-ER-cyt.pr-m]} = \frac{1}{4}\left\{ \frac{3075 \cdot Ca_{cyt}^2}{25 + Ca_{cyt}^2} + (Ca_{ER} - Ca_{cyt}) \right.$$

$$\left. \cdot \frac{51250 \cdot Ca_{cyt}}{(25 + Ca_{cyt}^2)^2} - 0.4 \cdot Ca_{cyt} - 5.0025 \right\}^2 + \frac{1}{4}\left\{ \frac{1562.5 \cdot Ca_m \cdot Ca_{cyt}}{(25 + Ca_{cyt}^2)^2} \right.$$

$$\left. - \frac{100.663296 \cdot Ca_{cyt}^7}{(0.16777216 + Ca_{cyt}^8)^2} + \frac{125 \cdot Ca_{cyt}^2}{25 + Ca_{cyt}^2} - 0.4 \cdot Ca_{cyt} - 0.03375 \right\}^2,$$

(ii) *Biological Yang-Mills energy of* **chaos** *calcium oscillations in the model involving endoplasmic reticulum, mitochondria, and cytosolic proteins.*

$$
\mathcal{EYM}^{\text{chaos}}_{[Ca^{2+}-ER-cyt.pr-m]} = \frac{1}{4} \left\{ \frac{0.75 \cdot k_{ch} \cdot Ca^2_{cyt}}{25 + Ca^2_{cyt}} + (Ca_{ER} - Ca_{cyt}) \right.
$$

$$
\cdot \frac{12.5 \cdot k_{ch} \cdot Ca_{cyt}}{(25 + Ca^2_{cyt})^2} - 0.4 \cdot Ca_{cyt} - 5.0025 \Bigg\}^2 + \frac{1}{4} \left\{ \frac{1562.5 \cdot Ca_m \cdot Ca_{cyt}}{(25 + Ca^2_{cyt})^2} \right.
$$

$$
\left. - \frac{100.663296 \cdot Ca^7_{cyt}}{(0.16777216 + Ca^8_{cyt})^2} + \frac{125 \cdot Ca^2_{cyt}}{25 + Ca^2_{cyt}} - 0.4 \cdot Ca_{cyt} - 0.03375 \right\}^2,
$$

where $k_{ch} \in [2780, 2980] \cup [3598, 3636]$;

(iii) *Biological Yang-Mills energy of* **birhythmicity** *calcium oscillations in the model involving endoplasmic reticulum, mitochondria, and cytosolic proteins.*

$$
\mathcal{EYM}^{\text{birhythmicity}}_{[Ca^{2+}-ER-cyt.pr-m]} = \frac{1}{4} \left\{ \frac{0.75 \cdot k_{ch} \cdot Ca^2_{cyt}}{25 + Ca^2_{cyt}} + (Ca_{ER} - Ca_{cyt}) \right.
$$

$$
\cdot \frac{12.5 \cdot k_{ch} \cdot Ca_{cyt}}{(25 + Ca^2_{cyt})^2} - 0.4 \cdot Ca_{cyt} - 5.0025 \Bigg\}^2 + \frac{1}{4} \left\{ \frac{1562.5 \cdot Ca_m \cdot Ca_{cyt}}{(25 + Ca^2_{cyt})^2} \right.
$$

$$
\left. - \frac{100.663296 \cdot Ca^7_{cyt}}{(0.16777216 + Ca^8_{cyt})^2} + \frac{125 \cdot Ca^2_{cyt}}{25 + Ca^2_{cyt}} - 0.4 \cdot Ca_{cyt} - 0.03375 \right\}^2,
$$

where $k_{ch} \in [1968, 2456]$.

13.3.3 Yang-Mills energetic surfaces of constant level. Theoretical biological interpretations

In the opinion of the authors of this book, from a biological point of view, the appearance in our geometrical studies of a biological electromagnetic field $\mathbb{F}$ (unknown until now), directly and naturally provided by a nonlinear ODEs system of first-order that governs some biological phenomena, may probably have interesting connections with the intrinsic biological phenomena studied. In this way, we believe that our geometric biological field $\mathbb{F}$ may be regarded as follows:

- Either this biological field $\mathbb{F}$ vanishes in order to realize a stability of the biological phenomena studied. This should be probably because such an "electromagnetic" field must not exist in such biological phenomena.

- Or this biological field $\mathbb{F}$ does not vanish, having a natural and microscopic character in the biological phenomena studied. In other words, this microscopic biological field $\mathbb{F}$ may be probably regarded as being provided not necessarily by the ODEs systems involved in the studies, but by the pair of Euclidian metrics $\Delta = (1, \delta_{ij})$, which have the well-known physical meaning of microscopic gravitational potentials produced intrinsically by the biological matter.

In this case, we believe that the classical geometric study (i.e., fundamental forms and main curvatures) of constant level Yang-Mills energy surfaces provided by the biological ODEs systems involved in this Chapter, aided by relevant computer imaging simulation, represents a promising research topic for Theoretical Biology.

For example, the geometric features of the 20-th Taylor approximation of the rational constant level Yang-Mills energy surfaces for the energies of Theorem 197 and of Theorem 202 respectively, which are represented below, exhibit specific geometric features which both reflect properties of the associated ODEs systems, and provide biological relevance for the modeled phenomena.

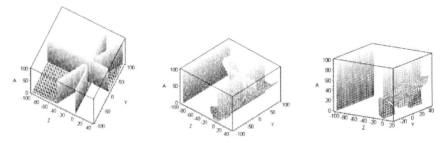

Fig. 1. Taylor approximation of rational constant level Yang-Mills energy surfaces for intracellular calcium oscillations (bursting, chaos, and quasiperiodicity cases).

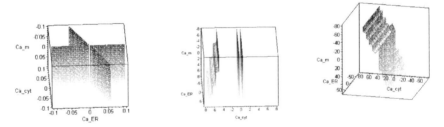

Fig. 2. Taylor approximation of rational constant level Yang-Mills energy surfaces for calcium oscillations through endoplasmic reticulum, mitochondria, and cytosolic proteins (bursting, chaos, and birhythmicity cases).

Open problem. Determine the biological meaning of our previously described geometrical Yang-Mills energies.

CHAPTER 14

JET GEOMETRICAL OBJECTS PRODUCED BY LINEAR ODEs SYSTEMS AND HIGHER-ORDER ODEs

Using the applicative geometrical theory from Chapter 10, in this Chapter we construct the Riemann-Lagrange geometry on 1-jet spaces (in the sense of d-connections, d-torsions, d-curvatures, electromagnetic d-fields, and geometric electromagnetic Yang-Mills energies), produced by a given linear ODEs system of first-order or by a given higher-order ODE. The case of a non-homogenous linear ODE of higher-order is also disscused.

14.1 JET RIEMANN-LAGRANGE GEOMETRY PRODUCED BY A NON-HOMOGENOUS LINEAR ODEs SYSTEM OF ORDER ONE

In this Section we apply our preceding jet Riemann-Lagrange geometrical results for a non-homogenous linear ODEs system of order one. In this direction, let us consider the following non-homogenous linear ODEs system of first-order, locally described, in a convenient chart on $J^1(\mathbb{R}, \mathbb{R}^n)$, by the differential equations

$$\frac{dx^i}{dt} = \sum_{k=1}^{n} a^{(i)}_{(1)k}(t)x^k + f^{(i)}_{(1)}(t), \ \forall \ i = \overline{1, n}, \tag{14.1}$$

Jet Single-Time Lagrange Geometry and Its Applications
1-st Edition. By Vladimir Balan and Mircea Neagu.
© 2011 John Wiley & Sons, Inc. Published 2011 John Wiley & Sons, Inc.

where the local components $a_{(1)k}^{(i)}$ and $f_{(1)}^{(i)}$ transform by the tensorial rules

$$a_{(1)k}^{(i)} = \frac{\partial x^i}{\partial \widetilde{x}^j} \frac{d\widetilde{t}}{dt} \cdot \widetilde{a}_{(1)k}^{(j)}, \ \forall \ k = \overline{1, n},$$

and

$$f_{(1)}^{(i)} = \frac{\partial x^i}{\partial \widetilde{x}^j} \frac{d\widetilde{t}}{dt} \cdot \widetilde{f}_{(1)}^{(j)}.$$

Remark 203. We suppose that the product manifold $\mathbb{R} \times \mathbb{R}^n \subset J^1(\mathbb{R}, \mathbb{R}^n)$ is endowed *a priori* with the pair of Euclidian metrics $\Delta = (1, \delta_{ij})$, with respect to the coordinates (t, x^i).

It is obvious that the non-homogenous linear ODEs system (14.1) is a particular case of the jet first-order nonlinear ODEs system (10.5) for

$$X_{(1)}^{(i)}(t, x) = \sum_{k=1}^{n} a_{(1)k}^{(i)}(t) x^k + f_{(1)}^{(i)}(t), \ \forall \ i = \overline{1, n}. \tag{14.2}$$

In order to present the main jet Riemann-Lagrange geometrical objects that characterize the non-homogenous linear ODEs system (14.1), we use the matrix notation

$$A_{(1)} = \left(a_{(1)j}^{(i)}(t) \right)_{i,j=\overline{1,n}}.$$

In this context, applying our preceding jet geometrical Riemann-Lagrange theory to the non-homogenous linear ODEs system (14.1) and the pair of Euclidian metrics $\Delta = (1, \delta_{ij})$, we get (cf. Theorem 180 and Remark 182) the following:

Theorem 204. (i) *The* **canonical nonlinear connection** *on* $J^1(\mathbb{R}, \mathbb{R}^n)$, **produced by the non-homogenous linear ODEs system (14.1)**, *has the local components*

$$\hat{\Gamma} = \left(0, \hat{N}_{(1)j}^{(i)} \right),$$

where $\hat{N}_{(1)j}^{(i)}$ *are the entries of the skew-symmetric matrix*

$$\hat{N}_{(1)} = \left(\hat{N}_{(1)j}^{(i)} \right)_{i,j=\overline{1,n}} = -\frac{1}{2} \left[A_{(1)} - {}^T A_{(1)} \right].$$

(ii) *All adapted components of the* **canonical generalized Cartan connection** $C\hat{\Gamma}$, **produced by the non-homogenous linear ODEs system (14.1)**, *are zero.*

(iii) *The effective adapted components* $\hat{R}_{(1)1j}^{(i)}$ *of the* **torsion d-tensor** $\hat{T}$ *of the canonical generalized Cartan connection* $C\hat{\Gamma}$, **produced by the non-homogenous linear ODEs system (14.1)**, *are the entries of the skew-symmetric matrices*

$$\hat{R}_{(1)1} = \left(\hat{R}_{(1)1j}^{(i)} \right)_{i,j=\overline{1,n}} = \frac{1}{2} \left[\dot{A}_{(1)} - {}^T \dot{A}_{(1)} \right],$$

where

$$\dot{A}_{(1)} = \frac{d}{dt}\left[A_{(1)}\right].$$

(iv) *All adapted components of the* **curvature** *d-tensor* $\hat{\mathbf{R}}$ *of the canonical generalized Cartan connection* $C\hat{\Gamma}$, **produced by the non-homogenous linear ODEs system (14.1)**, *cancel.*

(v) *The geometric* **electromagnetic distinguished 2-form, produced by the non-homogenous linear ODEs system (14.1)**, *is given by*

$$\hat{\mathbb{F}} = \hat{F}_{(i)j}^{(1)}\delta y_1^i \wedge dx^j,$$

where

$$\delta y_1^i = dy_1^i - \frac{1}{2}\left[a_{(1)k}^{(i)} - a_{(1)i}^{(k)}\right]dx^k, \ \forall \ i = \overline{1,n},$$

and the adapted components $\hat{F}_{(i)j}^{(1)}$ *are the entries of the skew-symmetric matrix*

$$\hat{F}^{(1)} = \left(\hat{F}_{(i)j}^{(1)}\right)_{i,j=\overline{1,n}} = -\hat{N}_{(1)} = \frac{1}{2}\left[A_{(1)} - {}^TA_{(1)}\right],$$

that is, we have

$$\hat{F}_{(i)j}^{(1)} = \frac{1}{2}\left[a_{(1)j}^{(i)} - a_{(1)i}^{(j)}\right].$$

(vi) *The* **jet geometric Yang-Mills energy, produced by the non-homogenous linear ODEs system (14.1)**, *is given by the formula*

$$\mathcal{E}\mathcal{Y}\mathcal{M}^{\text{NHLODEs}}(t) = \frac{1}{4}\sum_{i=1}^{n-1}\sum_{j=i+1}^{n}\left[a_{(1)j}^{(i)} - a_{(1)i}^{(j)}\right]^2.$$

Proof: Using relations (14.2), we easily deduce that we have the Jacobian matrix

$$J\left(X_{(1)}\right) = A_{(1)},$$

where $X_{(1)} := \left(X_{(1)}^{(i)}\right)_{i=\overline{1,n}}$.

Consequently, applying Theorem 180 to the non-homogenous linear ODEs system (14.1), together with Remark 182, we obtain the required results.

Remark 205. The entire jet Riemann-Lagrange geometry produced by the non-homogenous linear ODEs system (14.1) does not depend on the non-homogeneity terms $f_{(1)}^{(i)}(t)$.

Remark 206. The jet geometric Yang-Mills energy, produced by the non-homogenous linear ODEs system (14.1), cancels if and only if the matrix $A_{(1)}$ is a symmetric one. In this case, the entire jet Riemann-Lagrange geometry produced by the non-homogenous linear ODEs system (14.1) becomes trivial, so it does not offer geometrical information about the initial system (14.1). However, it is important to note that, in this particular situation, we have the symmetry of the matrix $A_{(1)}$, which implies that the matrix $A_{(1)}$ is diagonalizable.

Remark 207. All jet *torsion* adapted components, *produced by a non-homogenous linear ODEs system with constant coefficients* $a_{(1)j}^{(i)}$, are zero.

14.2 JET RIEMANN-LAGRANGE GEOMETRY PRODUCED BY A HIGHER-ORDER ODE

Let us consider the higher-order ODE expressed by

$$y^{(n)}(t) = f(t, y(t), y'(t), ..., y^{(n-1)}(t)), \ n \geq 2, \tag{14.3}$$

where $y(t)$ is the unknown function, $y^{(k)}(t)$ is the derivative of order k of the unknown function $y(t)$ for each $k \in \{0, 1, ..., n\}$ and f is a given differentiable function depending on the distinct variables $t, y(t), y'(t), ..., y^{(n-1)}(t)$.

It is well known the fact that, using the notations

$$x^1 = y, \ x^2 = y', \ ..., \ x^n = y^{(n-1)},$$

the higher-order ODE (14.3) is equivalent with the nonlinear ODEs system of order one

$$\begin{cases} \dfrac{dx^1}{dt} = x^2, \\[2mm] \dfrac{dx^2}{dt} = x^3, \\[2mm] \quad \vdots \\[2mm] \dfrac{dx^{n-1}}{dt} = x^n, \\[2mm] \dfrac{dx^n}{dt} = f(t, x^1, x^2, ..., x^n). \end{cases} \tag{14.4}$$

But, the first-order nonlinear ODEs system (14.4) can be regarded, in a convenient local chart, as a particular case of the jet nonlinear ODEs system of order one (10.5), setting

$$X_{(1)}^{(1)}(t, x) = x^2, X_{(1)}^{(2)}(t, x) = x^3, ..., X_{(1)}^{(n-1)}(t, x) = x^n,$$
$$X_{(1)}^{(n)}(t, x) = f(t, x^1, x^2, ..., x^n), \tag{14.5}$$

where we suppose that the geometrical object $X = \left(X_{(1)}^{(i)}(t, x) \right)$ behaves like a d-tensor on $J^1(\mathbb{R}, \mathbb{R}^n)$.

Remark 208. We assume again that the product manifold $\mathbb{R} \times \mathbb{R}^n \subset J^1(\mathbb{R}, \mathbb{R}^n)$ is endowed *a priori* with the pair of Euclidian metrics $\Delta = (1, \delta_{ij})$, with respect to the coordinates (t, x^i).

Definition 209. Any geometrical object on $J^1(\mathbb{R}, \mathbb{R}^n)$, which is produced by the first-order nonlinear ODEs system (14.4), is called a *geometrical object produced by the higher-order ODE (14.3)*.

In this context, the Riemann-Lagrange geometrical behavior on the 1-jet space $J^1(\mathbb{R}, \mathbb{R}^n)$ of the higher-order ODE (14.3) is described in the following result (cf. Theorem 180 and Remark 182):

Theorem 210. (i) *The* **canonical nonlinear connection** *on* $J^1(\mathbb{R}, \mathbb{R}^n)$, **produced by the higher-order ODE (14.3)**, *has the local components*

$$\check{\Gamma} = \left(0, \check{N}^{(i)}_{(1)j}\right),$$

where $\check{N}^{(i)}_{(1)j}$ *are the entries of the skew-symmetric matrix* $\check{N}_{(1)} = \left(\check{N}^{(i)}_{(1)j}\right)_{i,j=\overline{1,n}} =$

$$= -\frac{1}{2}
\begin{pmatrix}
0 & 1 & 0 & \cdots & 0 & 0 & -\dfrac{\partial f}{\partial x^1} \\[2mm]
-1 & 0 & 1 & \cdots & 0 & 0 & -\dfrac{\partial f}{\partial x^2} \\[2mm]
0 & -1 & 0 & \cdots & 0 & 0 & -\dfrac{\partial f}{\partial x^3} \\[2mm]
\vdots & \vdots & \vdots & \ddots & \vdots & \vdots & \vdots \\[2mm]
0 & 0 & 0 & \cdots & 0 & 1 & -\dfrac{\partial f}{\partial x^{n-2}} \\[2mm]
0 & 0 & 0 & \cdots & -1 & 0 & 1 - \dfrac{\partial f}{\partial x^{n-1}} \\[2mm]
\dfrac{\partial f}{\partial x^1} & \dfrac{\partial f}{\partial x^2} & \dfrac{\partial f}{\partial x^3} & \cdots & \dfrac{\partial f}{\partial x^{n-2}} & -1 + \dfrac{\partial f}{\partial x^{n-1}} & 0
\end{pmatrix}.$$

(ii) *All adapted components of the* **canonical generalized Cartan connection** $C\check{\Gamma}$, **produced by the higher-order ODE (14.3)**, *are zero*.

(iii) *The effective adapted components of the* **torsion** *d-tensor* $\check{\mathbf{T}}$ *of the canonical generalized Cartan connection* $C\check{\Gamma}$, **produced by the higher-order ODE (14.3)**, *are the entries of the skew-symmetric matrices*

$$\check{R}_{(1)1} = \frac{1}{2}
\begin{pmatrix}
0 & 0 & \cdots & 0 & -\dfrac{\partial^2 f}{\partial t \partial x^1} \\[2mm]
0 & 0 & \cdots & 0 & -\dfrac{\partial^2 f}{\partial t \partial x^2} \\[2mm]
\vdots & \vdots & \ddots & \vdots & \vdots \\[2mm]
0 & 0 & \cdots & 0 & -\dfrac{\partial^2 f}{\partial t \partial x^{n-1}} \\[2mm]
\dfrac{\partial^2 f}{\partial t \partial x^1} & \dfrac{\partial^2 f}{\partial t \partial x^2} & \cdots & \dfrac{\partial^2 f}{\partial t \partial x^{n-1}} & 0
\end{pmatrix}$$

and

$$
\check{R}_{(1)k} = -\frac{1}{2}
\begin{pmatrix}
0 & 0 & \cdots & 0 & -\dfrac{\partial^2 f}{\partial x^k \partial x^1} \\[2ex]
0 & 0 & \cdots & 0 & -\dfrac{\partial^2 f}{\partial x^k \partial x^2} \\[2ex]
\vdots & \vdots & \ddots & \vdots & \vdots \\[2ex]
0 & 0 & \cdots & 0 & -\dfrac{\partial^2 f}{\partial x^k \partial x^{n-1}} \\[2ex]
\dfrac{\partial^2 f}{\partial x^k \partial x^1} & \dfrac{\partial^2 f}{\partial x^k \partial x^2} & \cdots & \dfrac{\partial^2 f}{\partial x^k \partial x^{n-1}} & 0
\end{pmatrix},
$$

where $k \in \{1, 2, ..., n\}$.

(iv) *All adapted components of the* **curvature** *d-tensor* $\check{R}$ *of the canonical generalized Cartan connection* $C\check{\Gamma}$, **produced by the higher-order ODE (14.3),** *cancel.*

(v) *The geometric* **electromagnetic distinguished 2-form, produced by the higher-order ODE (14.3),** *has the form*

$$
\mathbb{F} = \check{F}^{(1)}_{(i)j} \delta y_1^i \wedge dx^j,
$$

where

$$
\delta y_1^i = dy_1^i + \check{N}^{(i)}_{(1)k} dx^k, \ \forall \, i = \overline{1, n},
$$

and the adapted components $\check{F}^{(1)}_{(i)j}$ *are the entries of the skew-symmetric matrix*

$$
\check{F}^{(1)} = \left(\check{F}^{(1)}_{(i)j} \right)_{i,j=\overline{1,n}} = -\check{N}_{(1)}.
$$

(vi) *The* **jet geometric Yang-Mills energy, produced by the higher-order ODE (14.3),** *is given by the formula*

$$
\mathcal{EYM}^{\text{HODE}}(t, x) = \frac{1}{4} \left[n - 1 - 2\frac{\partial f}{\partial x^{n-1}} + \sum_{j=1}^{n-1} \left(\frac{\partial f}{\partial x^j} \right)^2 \right].
$$

Proof: By partial derivation, relations (14.5) lead to the Jacobian matrix

$$
J\left(X_{(1)} \right) =
\begin{pmatrix}
0 & 1 & 0 & \cdots & 0 & 0 \\
0 & 0 & 1 & \cdots & 0 & 0 \\
\vdots & \vdots & \vdots & \ddots & \vdots & \vdots \\
0 & 0 & 0 & \cdots & 0 & 1 \\
\dfrac{\partial f}{\partial x^1} & \dfrac{\partial f}{\partial x^2} & \dfrac{\partial f}{\partial x^3} & \cdots & \dfrac{\partial f}{\partial x^{n-1}} & \dfrac{\partial f}{\partial x^n}
\end{pmatrix},
$$

where $X_{(1)} := \left(X_{(1)}^{(i)} \right)_{i=\overline{1,n}}$.

In conclusion, Theorem 180, together with Remark 182, applied to first-order nonlinear ODEs system (14.4), give what we were looking for.

14.3 RIEMANN-LAGRANGE GEOMETRY PRODUCED BY A NON-HOMOGENOUS LINEAR ODE OF HIGHER-ORDER

If we consider the non-homogenous linear ODE of order $n \in \mathbb{N}, n \geq 2$, expressed by

$$a_0(t)y^{(n)} + a_1(t)y^{(n-1)} + ... + a_{n-1}(t)y' + a_n(t)y = b(t), \qquad (14.6)$$

where $b(t)$ and $a_i(t), \forall\, i = \overline{0, n}$, are given differentiable real functions and $a_0(t) \neq 0$, $\forall\, t \in [a, b]$, then we recover the higher-order ODE (14.3) for the particular function

$$f(t, x) = \frac{b(t)}{a_0(t)} - \frac{a_n(t)}{a_0(t)} \cdot x^1 - \frac{a_{n-1}(t)}{a_0(t)} \cdot x^2 - ... - \frac{a_1(t)}{a_0(t)} \cdot x^n, \qquad (14.7)$$

where we recall that we have

$$y = x^1, y' = x^2, ..., y^{(n-1)} = x^n.$$

Consequently, we can derive the jet Riemann-Lagrange geometry attached to the non-homogenous linear higher-order ODE (14.6) (cf. Theorem 210).

Corollary 211. (i) *The* **canonical nonlinear connection** *on* $J^1(\mathbb{R}, \mathbb{R}^n)$, **produced by the non-homogenous linear higher-order ODE (14.6)**, *has the local components*

$$\tilde{\Gamma} = \left(0, \tilde{N}_{(1)j}^{(i)} \right),$$

where $\tilde{N}_{(1)j}^{(i)}$ *are the entries of the skew-symmetric matrix*

$$\tilde{N}_{(1)} = \left(\tilde{N}_{(1)j}^{(i)} \right)_{i,j=\overline{1,n}}$$

$$= -\frac{1}{2} \begin{pmatrix} 0 & 1 & 0 & \cdots & 0 & 0 & \dfrac{a_n}{a_0} \\ -1 & 0 & 1 & \cdots & 0 & 0 & \dfrac{a_{n-1}}{a_0} \\ 0 & -1 & 0 & \cdots & 0 & 0 & \dfrac{a_{n-2}}{a_0} \\ \vdots & \vdots & \vdots & \ddots & \vdots & \vdots & \vdots \\ 0 & 0 & 0 & \cdots & 0 & 1 & \dfrac{a_3}{a_0} \\ 0 & 0 & 0 & \cdots & -1 & 0 & 1+\dfrac{a_2}{a_0} \\ -\dfrac{a_n}{a_0} & -\dfrac{a_{n-1}}{a_0} & -\dfrac{a_{n-2}}{a_0} & \cdots & -\dfrac{a_3}{a_0} & -1-\dfrac{a_2}{a_0} & 0 \end{pmatrix}.$$

(ii) *All adapted components of the* **canonical generalized Cartan connection** $C\tilde{\Gamma}$, **produced by the non-homogenous linear higher-order ODE (14.6)**, *are zero.*

(iii) *All adapted components of the* **torsion** *d-tensor* $\tilde{\mathbf{T}}$ *of the canonical generalized Cartan connection* $C\tilde{\Gamma}$, **produced by the non-homogenous linear higher-order ODE (14.6)**, *are zero, except the temporal components*

$$\tilde{R}^{(i)}_{(1)1n} = -\tilde{R}^{(n)}_{(1)1i} = \frac{a'_{n-i+1}a_0 - a_{n-i+1}a'_0}{2a_0^2}, \ \forall \, i = \overline{1, n-1},$$

where we denoted by " ' " *the derivatives of the functions* $a_k(t)$.

(iv) *All adapted components of the* **curvature** *d-tensor* $\tilde{\mathbf{R}}$ *of the canonical generalized Cartan connection* $C\tilde{\Gamma}$, **produced by the non-homogenous linear higher-order ODE (14.6)**, *cancel.*

(v) *The geometric* **electromagnetic distinguished 2-form, produced by the non-homogenous linear higher-order ODE (14.6)**, *has the expression*

$$\tilde{\mathbb{F}} = \tilde{F}^{(1)}_{(i)j}\delta y_1^i \wedge dx^j,$$

where

$$\delta y_1^i = dy_1^i + \tilde{N}^{(i)}_{(1)k}dx^k, \ \forall \, i = \overline{1, n},$$

and the adapted components $\tilde{F}^{(1)}_{(i)j}$ *are the entries of the skew-symmetric matrix*

$$\tilde{F}^{(1)} = \left(\tilde{F}^{(1)}_{(i)j} \right)_{i,j=\overline{1,n}} = -\tilde{N}_{(1)}.$$

(vi) *The* **jet geometric Yang-Mills electromagnetic energy, produced by the non-homogenous linear higher-order ODE (14.6)**, *has the form*

$$\mathcal{EYM}^{\mathrm{NHLHODE}}(t) = \frac{1}{4}\left[n - 1 + 2\frac{a_2}{a_0} + \sum_{j=2}^{n} \frac{a_j^2}{a_0^2} \right].$$

Proof: We apply the Theorem 210 for the particular function (14.7) and we use the relations

$$\frac{\partial f}{\partial x^j} = -\frac{a_{n-j+1}}{a_0}, \ \forall \, j = \overline{1, n}.$$

Remark 212. The entire jet Riemann-Lagrange geometry produced by the non-homogenous linear higher-order ODE (14.6) is independent by the term of non-homogeneity $b(t)$. In the opinion of the authors, this fact emphasizes that the most important role in the study of the ODE (14.6) is played by its attached homogenous linear higher-order ODE.

■ **EXAMPLE 14.1**

The law of motion without friction of a material point of mass $m > 0$, which is placed on a spring having the constant of elasticity $k > 0$, is given by the homogenous linear ODE of order two (*harmonic oscillator*)

$$\frac{d^2y}{dt^2} + \omega^2 y = 0, \tag{14.8}$$

where the coordinate y measures the distance from the center of mass and $\omega^2 = k/m$. It follows that we have

$$n = 2, \quad a_0(t) = 1, \quad a_1(t) = 0 \quad \text{and} \quad a_2(t) = \omega^2,$$

that is, the second-order ODE (14.8) provides the following *jet geometric Yang-Mills electromagnetic energy of harmonic oscillator*:

$$\mathcal{EYM}^{\text{Harmonic Oscillator}} = \frac{1}{4}\left(1 + \omega^2\right)^2.$$

Open problem. Is there a real physical interpretation for the jet geometric Yang-Mills electromagnetic energy attached to the harmonic oscillator?

CHAPTER 15

JET SINGLE-TIME GEOMETRICAL EXTENSION OF THE KCC-INVARIANTS

In this Chapter we generalize on the 1-jet space $J^1(\mathbb{R}, M)$, where M^n is an arbitrary smooth manifold of dimension n, the basics of the KCC-theory for second-order systems of differential equations (see Kosambi [44], Cartan [26] and Chern [27]). In this respect, let us consider on $J^1(\mathbb{R}, M)$ a second-order system of differential equations (SODEs) of local form

$$\frac{d^2x^i}{dt^2} + F^{(i)}_{(1)1}(t, x^k, y^k_1) = 0, \quad i = \overline{1, n}, \tag{15.1}$$

where $y^k_1 = dx^k/dt$ and the local components $F^{(i)}_{(1)1}(t, x^k, y^k_1)$ transform, under a change of coordinates (1.2), by the rules

$$\widetilde{F}^{(r)}_{(1)1} = F^{(j)}_{(1)1} \left(\frac{dt}{d\widetilde{t}}\right)^2 \frac{\partial \widetilde{x}^r}{\partial x^j} - \frac{dt}{d\widetilde{t}} \frac{\partial \widetilde{y}^r_1}{\partial t} - \frac{\partial x^m}{\partial \widetilde{x}^j} \frac{\partial \widetilde{y}^r_1}{\partial x^m} \widetilde{y}^j_1. \tag{15.2}$$

Remark 213. The second-order system of differential equations (15.1) is invariant under a change of coordinates (1.2).

Using a temporal Riemannian metric $h_{11}(t)$ on $\mathbb{R}$ and taking into account the transformation rules (1.7) and (1.10), we can write the SODEs (15.1) in the following form:

$$\frac{d^2 x^i}{dt^2} - \kappa_{11}^1 y_1^i + 2G_{(1)1}^{(i)}(t, x^k, y_1^k) = 0, \quad i = \overline{1, n},$$

where

$$G_{(1)1}^{(i)} = \frac{1}{2} F_{(1)1}^{(i)} + \frac{1}{2} \kappa_{11}^1 y_1^i$$

are the components of a spatial semispray on $J^1(\mathbb{R}, M)$. Moreover, the coefficients of the spatial semispray $G_{(1)1}^{(i)}$ produce the spatial components $N_{(1)j}^{(i)}$ of a nonlinear connection Γ on the 1-jet space $J^1(\mathbb{R}, M)$ by putting

$$N_{(1)j}^{(i)} = \frac{\partial G_{(1)1}^{(i)}}{\partial y_1^j} = \frac{1}{2} \frac{\partial F_{(1)1}^{(i)}}{\partial y_1^j} + \frac{1}{2} \kappa_{11}^1 \delta_j^i.$$

In order to find the basic jet differential geometrical invariants of the system (15.1) (see the works [2], [4], [5], [19], [87]) under the jet coordinate transformations (1.2), we define the *h-KCC-covariant derivative of a d-tensor of kind* $T_{(1)}^{(i)}(t, x^k, y_1^k)$ on the 1-jet space $J^1(\mathbb{R}, M)$, via

$$
\begin{aligned}
\frac{\overset{h}{D} T_{(1)}^{(i)}}{dt} &= \frac{\partial T_{(1)}^{(i)}}{\partial t} + N_{(1)r}^{(i)} T_{(1)}^{(r)} - \kappa_{11}^1 T_{(1)}^{(i)} \\
&= \frac{\partial T_{(1)}^{(i)}}{\partial t} + \frac{1}{2} \frac{\partial F_{(1)1}^{(i)}}{\partial y_1^r} T_{(1)}^{(r)} - \frac{1}{2} \kappa_{11}^1 T_{(1)}^{(i)},
\end{aligned}
$$

where the Einstein summation convention is used throughout.

Remark 214. The *h-KCC-covariant derivative* components $\dfrac{\overset{h}{D} T_{(1)}^{(i)}}{dt}$ transform, under a change of coordinates (1.2), as a d-tensor of type $T_{(1)1}^{(i)}$.

In such a geometrical context, if we use the notation $y_1^i = dx^i/dt$, then the system (15.1) can be written in the following distinguished tensorial form:

$$
\begin{aligned}
\frac{\overset{h}{D} y_1^i}{dt} &= -F_{(1)1}^{(i)}(t, x^k, y_1^k) + N_{(1)r}^{(i)} y_1^r - \kappa_{11}^1 y_1^i \\
&= -F_{(1)1}^{(i)} + \frac{1}{2} \frac{\partial F_{(1)1}^{(i)}}{\partial y_1^r} y_1^r - \frac{1}{2} \kappa_{11}^1 y_1^i,
\end{aligned}
$$

Definition 215. The distinguished tensor

$$\overset{h(i)}{\varepsilon}_{(1)1} = -F_{(1)1}^{(i)} + \frac{1}{2} \frac{\partial F_{(1)1}^{(i)}}{\partial y_1^r} y_1^r - \frac{1}{2} \kappa_{11}^1 y_1^i$$

is called the *first h-KCC-invariant* on the 1-jet space $J^1(\mathbb{R}, M)$ of the SODEs (15.1), which is interpreted as an *external force* [2], [19].

■ **EXAMPLE 15.1**

It can be seen easily that for the particular first-order jet time-dependent dynamical system

$$\frac{dx^i}{dt} = X^{(i)}_{(1)}(t, x^k) \Rightarrow \frac{d^2 x^i}{dt^2} = \frac{\partial X^{(i)}_{(1)}}{\partial t} + \frac{\partial X^{(i)}_{(1)}}{\partial x^m} y_1^m, \tag{15.3}$$

where $X^{(i)}_{(1)}(t, x)$ is a given d-tensor on $J^1(\mathbb{R}, M)$, the first h-KCC-invariant has the form

$$\overset{h(i)}{\varepsilon}^{(i)}_{(1)1} = \frac{\partial X^{(i)}_{(1)}}{\partial t} + \frac{1}{2} \frac{\partial X^{(i)}_{(1)}}{\partial x^r} y_1^r - \frac{1}{2} \kappa^1_{11} y_1^i.$$

In the sequel, let us vary the trajectories $x^i(t)$ of the system (15.1) by the nearby trajectories $(\overline{x}^i(t, s))_{s \in (-\varepsilon, \varepsilon)}$, where $\overline{x}^i(t, 0) = x^i(t)$. Then, considering the *variation d-tensor field*

$$\xi^i(t) = \left. \frac{\partial \overline{x}^i}{\partial s} \right|_{s=0},$$

we get the *variational equations*

$$\frac{d^2 \xi^i}{dt^2} + \frac{\partial F^{(i)}_{(1)1}}{\partial x^j} \xi^j + \frac{\partial F^{(i)}_{(1)1}}{\partial y_1^r} \frac{d\xi^r}{dt} = 0. \tag{15.4}$$

In order to find other jet geometrical invariants for the system (15.1), we also introduce the *h-KCC-covariant derivative of a d-tensor of kind $\xi^i(t)$* on the 1-jet space $J^1(\mathbb{R}, M)$, via

$$\frac{\overset{h}{D}\xi^i}{dt} = \frac{d\xi^i}{dt} + N^{(i)}_{(1)m} \xi^m = \frac{d\xi^i}{dt} + \frac{1}{2} \frac{\partial F^{(i)}_{(1)1}}{\partial y_1^m} \xi^m + \frac{1}{2} \kappa^1_{11} \xi^i.$$

Remark 216. The *h-KCC-covariant derivative* components $\frac{\overset{h}{D}\xi^i}{dt}$ transform, under a change of coordinates (1.2), as a d-tensor $T^{(i)}_{(1)}$.

In this geometrical context, the variational equations (15.4) can be written in the following distinguished tensorial form:

$$\frac{\overset{h}{D}}{dt}\left(\frac{\overset{h}{D}\xi^i}{dt}\right) = \overset{h}{P}^i_{m11} \xi^m,$$

where

$$\overset{h}{P}^i_{j11} = -\frac{\partial F^{(i)}_{(1)1}}{\partial x^j} + \frac{1}{2} \frac{\partial^2 F^{(i)}_{(1)1}}{\partial t \partial y_1^j} + \frac{1}{2} \frac{\partial^2 F^{(i)}_{(1)1}}{\partial x^r \partial y_1^j} y_1^r - \frac{1}{2} \frac{\partial^2 F^{(i)}_{(1)1}}{\partial y_1^r \partial y_1^j} F^{(r)}_{(1)1}$$

$$+ \frac{1}{4} \frac{\partial F^{(i)}_{(1)1}}{\partial y_1^r} \frac{\partial F^{(r)}_{(1)1}}{\partial y_1^j} + \frac{1}{2} \frac{d\kappa^1_{11}}{dt} \delta^i_j - \frac{1}{4} \kappa^1_{11} \kappa^1_{11} \delta^i_j.$$

Definition 217. The d-tensor $\overset{h}{P^i_{j11}}$ is called the *second h-KCC-invariant* on the 1-jet space $J^1(\mathbb{R}, M)$ of the system (15.1), or the *jet h-deviation curvature d-tensor*.

■ **EXAMPLE 15.2**

If we consider the second-order system of differential equations of the *harmonic curves associated to the pair of Riemannian metrics* $(h_{11}(t), \varphi_{ij}(x))$, the system which is given by (see Example 1.8)

$$\frac{d^2x^i}{dt^2} - \kappa^1_{11}(t)\frac{dx^i}{dt} + \gamma^i_{jk}(x)\frac{dx^j}{dt}\frac{dx^k}{dt} = 0,$$

where $\kappa^1_{11}(t)$ and $\gamma^i_{jk}(x)$ are the Christoffel symbols of the Riemannian metrics $h_{11}(t)$ and $\varphi_{ij}(x)$, then the second h-KCC-invariant has the form

$$\overset{h}{P^i_{j11}} = -\mathfrak{R}^i_{pqj}y^p_1y^q_1,$$

where

$$\mathfrak{R}^i_{pqj} = \frac{\partial\gamma^i_{pq}}{\partial x^j} - \frac{\partial\gamma^i_{pj}}{\partial x^q} + \gamma^r_{pq}\gamma^i_{rj} - \gamma^r_{pj}\gamma^i_{rq}$$

are the components of the curvature of the spatial Riemannian metric $\varphi_{ij}(x)$. Consequently, the variational equations (15.4) become the following *jet Jacobi field equations*:

$$\frac{\overset{h}{D}}{dt}\left(\frac{\overset{h}{D\xi^i}}{dt}\right) + \mathfrak{R}^i_{pqm}y^p_1y^q_1\xi^m = 0,$$

where

$$\frac{\overset{h}{D\xi^i}}{dt} = \frac{d\xi^i}{dt} + \gamma^i_{jm}y^j_1\xi^m.$$

■ **EXAMPLE 15.3**

For the particular first-order jet time-dependent dynamical system (15.3), the jet h-deviation curvature d-tensor is given by

$$\overset{h}{P^i_{j11}} = \frac{1}{2}\frac{\partial^2 X^{(i)}_{(1)}}{\partial t\partial x^j} + \frac{1}{2}\frac{\partial^2 X^{(i)}_{(1)}}{\partial x^j\partial x^r}y^r_1 + \frac{1}{4}\frac{\partial X^{(i)}_{(1)}}{\partial x^r}\frac{\partial X^{(r)}_{(1)}}{\partial x^j} + \frac{1}{2}\frac{d\kappa^1_{11}}{dt}\delta^i_j - \frac{1}{4}\kappa^1_{11}\kappa^1_{11}\delta^i_j.$$

Definition 218. The distinguished tensors

$$\overset{h}{R^i_{jk1}} = \frac{1}{3}\left(\frac{\partial\overset{h}{P^i_{j11}}}{\partial y^k_1} - \frac{\partial\overset{h}{P^i_{k11}}}{\partial y^j_1}\right), \qquad \overset{h}{B^i_{jkm}} = \frac{\partial\overset{h}{R^i_{jk1}}}{\partial y^m_1},$$

and

$$D^{i1}_{jkm} = \frac{\partial^3 F^{(i)}_{(1)1}}{\partial y^j_1 \partial y^k_1 \partial y^m_1}$$

are called the *third, fourth,* and *fifth h-KCC-invariant* on the 1-jet vector bundle $J^1(\mathbb{R}, M)$ of the system (15.1).

Remark 219. Taking into account the transformation rules (15.2) of the components $F^{(i)}_{(1)1}$, we immediately deduce that the components D^{i1}_{jkm} behave like a d-tensor.

■ **EXAMPLE 15.4**

For the first-order jet time-dependent dynamical system (15.3) the third, fourth and fifth h-KCC-invariants are zero.

Theorem 220 (of characterization of the jet h-KCC-invariants). *All the five h-KCC-invariants of the system (15.1) cancel on $J^1(\mathbb{R}, M)$ if and only if there exists a flat symmetric linear connection $\Gamma^i_{jk}(x)$ on M such that*

$$F^{(i)}_{(1)1} = \Gamma^i_{pq}(x)y^p_1 y^q_1 - \kappa^1_{11}(t)y^i_1. \tag{15.5}$$

Proof: "⇐" By a direct calculation, we obtain

$$\overset{h(i)}{\varepsilon}_{(1)1} = 0, \quad \overset{h}{P}^i_{j11} = -\mathcal{R}^i_{pqj}y^p_1 y^q_1 = 0 \text{ and } D^{i1}_{jkl} = 0,$$

where $\mathcal{R}^i_{pqj} = 0$ are the components of the curvature of the flat symmetric linear connection $\Gamma^i_{jk}(x)$ on M.
"⇒" By integration, the relation

$$D^{i1}_{jkl} = \frac{\partial^3 F^{(i)}_{(1)1}}{\partial y^j_1 \partial y^k_1 \partial y^l_1} = 0$$

subsequently leads to

$$\frac{\partial^2 F^{(i)}_{(1)1}}{\partial y^j_1 \partial y^k_1} = 2\Gamma^i_{jk}(t, x) \Rightarrow \frac{\partial F^{(i)}_{(1)1}}{\partial y^j_1} = 2\Gamma^i_{jp}y^p_1 + \mathcal{U}^{(i)}_{(1)j}(t, x)$$

$$\Rightarrow F^{(i)}_{(1)1} = \Gamma^i_{pq}y^p_1 y^q_1 + \mathcal{U}^{(i)}_{(1)p}y^p_1 + \mathcal{V}^{(i)}_{(1)1}(t, x),$$

where the local functions $\Gamma^i_{jk}(t, x)$ are symmetrical in the indices j and k.
The equality $\overset{h(i)}{\varepsilon}_{(1)1} = 0$ on $J^1(\mathbb{R}, M)$ infers

$$\mathcal{V}^{(i)}_{(1)1} = 0, \quad \mathcal{U}^{(i)}_{(1)j} = -\kappa^1_{11}\delta^i_j.$$

Consequently, we have

$$\frac{\partial F^{(i)}_{(1)1}}{\partial y^j_1} = 2\Gamma^i_{jp}y^p_1 - \kappa^1_{11}\delta^i_j$$

and

$$F^{(i)}_{(1)1} = \Gamma^i_{pq} y^p_1 y^q_1 - \kappa^1_{11} y^i_1.$$

The condition $\overset{h}{P}{}^i_{j11} = 0$ on $J^1(\mathbb{R}, M)$ implies the equalities $\Gamma^i_{jk} = \Gamma^i_{jk}(x)$ and

$$\mathcal{R}^i_{pqj} + \mathcal{R}^i_{qpj} = 0,$$

where

$$\mathcal{R}^i_{pqj} = \frac{\partial \Gamma^i_{pq}}{\partial x^j} - \frac{\partial \Gamma^i_{pj}}{\partial x^q} + \Gamma^r_{pq}\Gamma^i_{rj} - \Gamma^r_{pj}\Gamma^i_{rq}.$$

It is important to note that, taking into account the transformation laws (1.2), (15.2), (1.7), and Example 1.5, we deduce that the local coefficients $\Gamma^i_{jk}(x)$ locally transform like a symmetric linear connection on M. Consequently, $\mathcal{R}^i_{pqj}$ represent the curvature of this symmetric linear connection.

On the other hand, the equality $\overset{h}{R}{}^i_{jk1} = 0$ leads us to $\mathcal{R}^i_{qjk} = 0$, which infers that the symmetric linear connection $\Gamma^i_{jk}(x)$ on M is flat.

REFERENCES

1. M. Anastasiei, H. Kawaguchi, *A geometrical theory of time-dependent Lagrangians, I. Nonlinear connections, II. M-Connections*, Tensor N.S. **48** (1989), 273–282, 283–293.

2. P. L. Antonelli, *Equivalence Problem for Systems of Second Order Ordinary Differential Equations*, Encyclopedia of Mathematics, Kluwer Academic Publishers, Dordrecht, 2000.

3. P. L. Antonelli, L. Bevilacqua, S. F. Rutz, *Theories and models in symbiogenesis*, Nonlinear Analysis: Real World Applications **4** (2003), 743–753.

4. P. L. Antonelli, I. Bucătaru, *New results about the geometric invariants in KCC-theory*, An. Şt. Univ. "Al. I. Cuza" Iaşi. Mat. N.S. **47** (2001), 405–420.

5. P. L. Antonelli, R. S. Ingarden, M. Matsumoto, *The Theory of Sprays and Finsler Spaces with Applications in Physics and Biology*, Fundamental Theories of Physics, vol. **58**, Kluwer Academic Publishers, Dordrecht, 1993.

6. P. L. Antonelli, R. Miron, *Lagrange and Finsler Geometry. Applications to Physics and Biology*, Kluwer Academic Publishers, 1996.

7. G. S. Asanov, *Finslerian Extension of General Relativity*, Reidel, Dordrecht, 1984.

8. G. S. Asanov, *Jet extension of Finslerian gauge approach*, Fortschritte der Physik **38**, no. **8** (1990), 571–610.

9. Gh. Atanasiu, V. Balan, N. Brînzei, M. Rahula, *The Differential-Geometric Structures: Tangent Bundles, Connections in Bundles and The Exponential Law in Jet-Spaces*, LKI, Moscow, 2010 (in Russian).

Jet Single-Time Lagrange Geometry and Its Applications   **185**
1-st Edition. By Vladimir Balan and Mircea Neagu.
© 2011 John Wiley & Sons, Inc. Published 2011 John Wiley & Sons, Inc.

10. Gh. Atanasiu, V. Balan, M. Neagu, *The Pavlov's 4-polyform of momenta* $K(p) = \sqrt[4]{p_1 p_2 p_3 p_4}$ *and its applications in Hamilton geometry*, Hypercomplex Numbers in Geometry and Physics **2** (**4**), vol. **2** (2005), 134–139 (in Russian).

11. Gh. Atanasiu, M. Neagu, *Jet local Riemann-Finsler geometry for the three-dimensional time*, Hypercomplex Numbers in Geometry and Physics **2** (**14**), vol. **7** (2010), in press.

12. Gh. Atanasiu, M. Neagu, *On Cartan spaces with the m-th root metric* $K(x, p) = \sqrt[m]{a^{i_1 i_2 \cdots i_m}(x) p_{i_1} p_{i_2} \cdots p_{i_m}}$, Hypercomplex Numbers in Geometry and Physics **2** (**12**), vol. **6** (2009), 67–73.

13. V. Balan, *Generalized Maxwell and Lorentz equations on first-order geometrized jet spaces*, Proceeding of the International Conference in Geometry and Topology, Cluj-Napoca, Romania, 1-5 October 2002; University of Cluj-Napoca Editors (2004), 11–24.

14. V. Balan, *Notable curves in geometrized* $J^1(T, M)$ *framework*, Balkan Journal of Geometry and Its Applications, vol. **8**, no. **2** (2003), 1–10.

15. V. Balan, *Numerical multilinear algebra of symmetric m-root structures. Spectral properties and applications*, In "Symmetry: Culture and Science," Symmetry Festival 2009, "Symmetry in the History of Science, Art and Technology; Part **2**: Geometric Approaches to Symmetry," 2010; Budapest, Hungary, **21** (**1–3**) (2010), 119–131.

16. V. Balan, M. Neagu, *Jet Finslerian geometry of the conformal Minkowski metric*, http://arXiv.org/math.DG/1102.4166v1 (2011).

17. V. Balan, M. Neagu, *Jet geometrical extension of the KCC-invariants*, Balkan Journal of Geometry and Its Applications, vol. **15**, no. **1** (2010), 8–16.

18. V. Balan, M. Neagu, *The 1-jet generalized Lagrange geometry induced by the rheonomic Chernov metric*, University Politehnica of Bucharest, Scientific Bulletin, series **A**, vol. **72**, iss. **3** (2010), 5–16.

19. V. Balan, I. R. Nicola, *Linear and structural stability of a cell division process model*, International Journal of Mathematics and Mathematical Sciences, vol. (2006), 1–15.

20. D. Bao, S. S. Chern, Z. Shen, *An Introduction to Riemann-Finsler Geometry*, Springer-Verlag, 2000.

21. P. Birtea, M. Puta, T. S. Raţiu, R. Tudoran, *A short proof of chaos in an atmospheric system*, http://xxx.lanl.gov/math.DS/0203155, (2002).

22. J. Benhabib, T. Miyao, *Some new results on the dynamics of the generalized Tobin model*, International Economic Review **22**, no. **3** (1981), 589–596.

23. J. A. M. Borghans, G. Dupont, A. Goldbeter, *Complex intracellular calcium oscillations: A theoretical exploration of possible mechanisms*, Biophys. Chem., **66** (1997), 25–41.

24. I. Bucătaru, R. Miron, *Finsler-Lagrange Geometry. Applications to Dynamical Systems*, Romanian Academy Eds, Bucharest, 2007.

25. J.F. Cariñena, J. de Lucas, *Lie systems: theory, generalisations, and applications*, arXiv.org/math-ph/1103.4166v1 (2011).

26. E. Cartan, *Observations sur le mémoir précédent*, Mathematische Zeitschrift **37**, no. **1** (1933), 619–622.

27. S.S. Chern, *Sur la géométrie d'un système d'equations differentialles du second ordre*, Bulletin des Sciences Mathématiques **63** (1939), 206–212.

28. V. M. Chernov, *On defining equations for the elements of associative and commutative algebras and on associated metric forms*, In: "Space-Time Structure. Algebra and Geometry" (D.G. Pavlov, Gh. Atanasiu, V. Balan Eds.), Russian Hypercomplex Society, Lilia Print, Moscow, 2007, 189–209.

29. M. Crampin, *Generalized Bianchi identities for horizontal distributions*, Math. Proc. Camb. Phil. Soc. **94** (1983), 125–132.

30. M. Crampin, F. A. E. Pirani, *Applicable Differential Geometry*, Cambridge University Press, 1986.

31. M. Crampin, G. E. Prince, G. Thompson, *A geometrical version of the Helmholtz conditions in time-dependent Lagrangian dynamics*, J. Phys. A: Math. Gen. **17** (1984), 1437–1447.

32. A. Das, A. DeBenedictis, S. Kloster, N. Tariq, *Relativistic particle, fluid and plasma mechanics coupled to gravity*, http://arXiv.org/math-ph/0503027v1 (2005).

33. J. Eells, L. Lemaire, *A report on harmonic maps*, Bull. London Math. Soc. **10** (1978), 1–68.

34. C. Frigioiu, *Noether's laws of conservation in time-dependent Lagrange spaces*, In: "Lagrange and Hamilton Geometries and Their Applications" (Radu Miron, Ed.), Fair Partners Publishers, Bucharest, 2004, 125–131.

35. G. Gandolfo, *Economic Dynamics*, Springer-Verlag, 1997.

36. G. I. Garas'ko, *Foundations of Finsler Geometry for Physicists*, Tetru Eds, Moscow, 2009 (in Russian).

37. G. I. Garas'ko, D.G. Pavlov, *The notions of distance and velocity modulus in the linear Finsler spaces*, In: "Space-Time Structure. Algebra and Geometry" (D.G. Pavlov, Gh. Atanasiu, V. Balan, Eds.); Russian Hypercomplex Society, Lilia Print, Moscow, 2007; 104–117.

38. A. L. Garner, Y. Y. Lau, D.W. Jordan, M. D. Uhler, R. M. Gilgenbach, *Implication of a simple mathematical model to cancer cell population dynamics*, Cell Prolif. **39** (2006), 15–28.

39. G. Giachetta, L. Mangiarotti, G. Sardanashvily, *Differential geometry of time-dependent mechanics*, http://arXiv.org/dg-ga/9702020v1 (1997).

40. V. Gîrțu, C. Ciubotariu, *Finsler spaces associated with a general relativistic magnetized plasma*, Tensor N. S. **56**, no. 1 (1995), 19–26.

41. M. Gîrțu, V. Gîrțu, M. Postolache, *Geodesic equations for magnetized plasma in GL-spaces*, BSG Proceedings **3**, Geometry Balkan Press, Bucharest, 1999, 27–34.

42. G. Houart, G. Dupont, A. Goldbeter, *Bursting, chaos and birhythmicity originating from self-modulation of the inositol* 1, 4, 5-*trisphosphate signal in model for intracellular* Ca^{2+} *oscillations*, Bulletin of Mathematical Biology **61** (1999), 507–530.

43. K. Kleidis, A. Kuiroukidis, D. Papadopoulos, L. Vlahos, *Magnetohydrodynamics and plasma cosmology*, http://arXiv.org/gr-qc/0512131v1 (2005).

44. D. D. Kosambi, *Parallelism and path-spaces*, Mathematische Zeitschrift **37**, no. **1** (1933), 608–618.

45. O. Krupková, *The Geometry of Ordinary Variational Equations*, Springer-Verlag, 1997.

46. O. Krupková, G. E. Prince, *Second-order ordinary differential equations in jet bundles and the inverse problem of the calculus of variations*, In: "Handbook of Global Analysis," Elsevier, 2008, 837–904.

47. M. de León, P. Rodrigues, *Generalized Classical Mechanics and Field Theory*, North-Holland, Amsterdam, 1985.

48. M. de León, P. R. Rodrigues, *Methods of Differential Geometry in Analytical Mechanics*, North-Holland, Amsterdam, 1989.

49. E. N. Lorenz, *On the existence of a slow manifold*, J. Atmos. Sci. **43** (1986), 1547-1557.

50. M. Marhl, T. Haberichter, M. Brumen, R. Heinrich, *Complex calcium oscillations and the role of mitochondria and cytosolic proteins*, Biosystems **57** (2000), 75–86.

51. E. Martínez, J. F. Cariñena, W. Sarlet, *Geometric characterization of separable second-order equations*, Math. Proc. Camb. Phil. Soc. **113** (1993), 205–224.

52. M. Matsumoto, *On Finsler spaces with curvature tensors of some special forms*, Tensor N.S. **22** (1971), 201–204.

53. M. Matsumoto, H. Shimada, *On Finsler spaces with 1-form metric. II. Berwald-Moór's metric $L = \left(y^1 y^2 ... y^n \right)^{1/n}$*, Tensor N.S. **32** (1978), 275–278.

54. R. V. Mikhailov, *On some questions of four dimensional topology: a survey of modern research*, Hypercomplex Numbers in Geometry and Physics **1** (**1**), vol. **1** (2004), 100–103.

55. R. Miron, M. Anastasiei, *The Geometry of Lagrange Spaces: Theory and Applications*, Kluwer Academic Publishers, 1994.

56. R. Miron, D. Hrimiuc, H. Shimada, S.V. Sabău, *The Geometry of Hamilton and Lagrange Spaces*, Kluwer Academic Publishers, 2001.

57. M. Neagu, *A relativistic approach on 1-jet spaces of the rheonomic Berwald-Moór metric*, Applied Sciences, vol. **13** (2011), 82–96.

58. M. Neagu, *Geometrical objects on the first-order jet space $J^1(T, \mathbb{R}^5)$ produced by the Lorenz atmospheric DEs system*, Carpathian Journal of Mathematics, vol. **26**, no. **2** (2010), 222-229.

59. M. Neagu, *Jet geometrical objects depending on a relativistic time*, An. Şt. Univ. "Al. I. Cuza" Iaşi (S.N.). Mat., Tomul **LVI**, f. **2** (2010), 407–428.

60. M. Neagu, *Jet geometrical objects produced by linear ODEs systems and superior order ODEs*, Studia Universitatis Babeş-Bolyai, Mathematica, vol. **LVI**, no. **1** (2011), 85–99.

61. M. Neagu, *Jet Riemann-Lagrange geometry applied to evolution DEs systems from economy*, Bulletin of Transilvania University of Braşov, no. **14** (**49**), series **B** (supplement) (2007), 199–210.

62. M. Neagu, *Local Bianchi identities in the relativistic non-autonomous Lagrange geometry*, http://arXiv.org/math.DG/0912.5209v1 (2009).

63. M. Neagu, *Ricci and Bianchi identities for h-normal Γ-linear connections on $J^1(T, M)$*, Hindawi Publishing Corporation, International Journal of Mathematics and Mathematical Sciences, no. **34** (2003), 2177–2192.

64. M. Neagu, *Riemann-Lagrange geometric dynamics for the multi-time magnetized non-viscous plasma*, http://arXiv.org/math.DG/1005.4567v1 (2010).

65. M. Neagu, *Riemann-Lagrange Geometry on 1-Jet Spaces*, Matrix Rom, Bucharest, 2005.

66. M. Neagu, *The geometry of autonomous metrical multi-time Lagrange space of electrodynamics*, International Journal of Mathematics and Mathematical Sciences, vol. **29**, no. **1** (2002), 7–15.

67. M. Neagu, *The geometry of relativistic rheonomic Lagrange spaces*, BSG Proceedings **5**, Geometry Balkan Press, Bucharest (2001), 142–168.

68. M. Neagu, I. R. Nicola, *Geometric dynamics of calcium oscillations ODEs systems*, Balkan Journal of Geometry and Its Applications, vol. **9**, no. **2** (2004), 36–67.

69. M. Neagu, E. Stoica, *Deflection d-tensor identities in the relativistic time-dependent Lagrange geometry*, Journal of Advanced Mathematical Studies, vol. **3**, no. **2** (2010), 71–88.

70. M. Neagu, C. Udriște, *A Riemann-Lagrange geometrization for metrical multi-time Lagrange spaces*, Balkan Journal of Geometry and Its Applications, vol. **11**, no. **1** (2006), 87–98.

71. M. Neagu, C. Udriște, *From PDEs systems and metrics to multi-time field theories and geometric dynamics*, Seminarul de Mecanică **79** (2001), Timișoara, Romania.

72. M. Neagu, C. Udriște, *Geometric dynamics of plasma in jet spaces with Berwald-Moór metric*, Proceedings of the 9-th WSEAS International Conference on System Science and Simulation in Engineering (ICOSSSE '10), Iwate, Japan, October 4–6, 2010 * selected topics in system science and simulation in engineering (2010), 42–49.

73. M. Neagu, C. Udriște, A. Oană, *Multi-time dependent sprays and h-traceless maps on $J^1(T, M)$*, Balkan Journal of Geometry and Its Applications, vol. **10**, no. **2** (2005), 76–92.

74. I. R. Nicola, *Geometric Methods for the Study of Some Biological Complex Systems*, Bren Eds, Bucharest, 2007 (in Romanian).

75. I. R. Nicola, M. Neagu, *Jet Riemann-Lagrange geometry and some applications in theoretical biology*, Journal of Dynamical Systems and Geometric Theories, vol. **6**, no. **1** (2008), 13–25.

76. V. Obădeanu, *Sisteme Dinamice Diferențiale. Dinamica Materiei Amorfe*, Editura Universității de Vest, Timișoara, Romania, 2006 (in Romanian).

77. P. J. Olver, *Applications of Lie Groups to Differential Equations*, Springer-Verlag, 1986.

78. D. B. Papadopoulos, *Plasma waves driven by gravitational waves in an expanding universe*, http://arXiv.org/gr-qc/0205096v1 (2002).

79. D. G. Pavlov, *Chronometry of three-dimensional time*, Hypercomplex Numbers in Geometry and Physics **1** (**1**), vol. **1** (2004), 19–30.

80. D. G. Pavlov, *Four-dimensional time*, Hypercomplex Numbers in Geometry and Physics **1** (**1**), vol. **1** (2004), 31–39.

81. D. G. Pavlov, *Generalization of scalar product axioms*, Hypercomplex Numbers in Geometry and Physics **1** (**1**), vol. **1** (2004), 5–18.

82. D. G. Pavlov, S. S. Kokarev, *Conformal gauges of the Berwald-Moór Geometry and their induced non-linear symmetries* (in Russian), Hypercomplex Numbers in Geometry and Physics **2** (**10**), vol. **5** (2008), 3-14.

83. A. S. Perelson, P. W. Nelson, *Mathematical analysis of HIV-1 dynamics in vivo*, Siam Review **41**, no. **1** (1999), 3-44.

84. H. Poincaré, *Sur les courbes definies par les equations différentielle*, C.R. Acad. Sci., Paris **90** (1880), 673–675.

85. H. Poincaré, *Sur une forme nouvelle des équations de la méchanique,* C.R. Acad Sci., Paris **132** (1901), 369–371.

86. B. Punsly, *Black Hole Gravitohydromagnetics*, Springer-Verlag, Berlin, 2001.

87. V. S. Sabău, *Systems biology and deviation curvature tensor*, Nonlinear Analysis. Real World Applications **6**, no. **3** (2005), 563–587.

88. W. Sarlet, *Geometrical structures related to second-order equations*, Proc. Conf. "Differential Geometry and Its Applications," Brno, Czechoslovakia, 1986 (D. Krupka and A. Švec Eds.), D. Reidel, Dordrecht (1986), 279–288.

89. D. J. Saunders, *The Geometry of Jet Bundles*, Cambridge University Press, New York, London, 1989.

90. H. Shimada, *On Finsler spaces with the metric*
$$L(x,y) = \sqrt[m]{a_{i_1 i_2 \ldots i_m}(x) y^{i_1} y^{i_2} \ldots y^{i_m}}, \text{ Tensor N. S. } \mathbf{33} \text{ (1979), 365–372.}$$

91. G. I. Solyanik, N. M. Berezetskaya, R. I. Bulkiewicz, G. I. Kulik, *Different growth patterns of a cancer cell population as a function of its starting growth characteristichs: Analysis by mathematical modeling*, Cell Prolif. **28** (1995), 263.

92. E. Sóos, *Géométrie et electromagnetisme*, Seminarul de Mecanică **35** (1992), Timişoara, Romania.

93. J. Tobin, *Money and economic growth*, Econometrica **33** (1965), 671-684.

94. C. Udrişte, *Geometric Dynamics*, Kluwer Academic Publishers, 2000.

95. C. Udrişte, M. Ferrara, D. Opriş, *Economic Geometric Dynamics*, Geometry Balkan Press, Bucharest, 2004.

96. S. Văcaru, *Interactions, Strings and Isotopies in Higher Order Anisotropic Superspaces*, Hadronic Press, 1998.

97. S. Văcaru, P. C. Stavrinos, E. Gaburov and D. Gontsa, *Clifford and Riemann-Finsler Structures in Geometric Mechanics and Gravity*, Geometry Balkan Press, 2006.

98. A. Vondra, *Sprays and homogeneous connections on* $\mathbb{R} \times TM$, Archivum Mathematicum (Brno), vol. **28** (1992), 163–173.

99. A. Vondra, *Symmetries of connections on fibered manifolds*, Archivum Mathematicum (Brno), vol. **30**, no. **2** (1994), 97–115.

100. N. M. Woods, K. S. R. Cuthbertson, P. H. Cobbold, *Repetitive transient rises in cytoplasmic free calcium in hormone-stimulated hepatocytes*, Nature **319** (1986), 600–602.

INDEX